Lecture Notes in Mathematics

Edited by A. Dold and B. Eckmann

1308

Peter Imkeller

Two-Parameter Martingales and Their Quadratic Variation

Springer-Verlag
Berlin Heidelberg New York London Paris Tokyo

Author

Peter Imkeller
Mathematisches Institut der Ludwig-Maximilians-Universität München
Theresienstr. 39, 8000 München, Federal Republic of Germany

Mathematics Subject Classification (1980): 60G07, 60G44, 60E15, 60G42, 60G48, 60G55, 60G60, 60H05

ISBN 978-3-540-19233-6 Springer-Verlag Berlin Heidelberg New York
ISBN 978-0-387-19233-8 Springer-Verlag New York Berlin Heidelberg

Printing and binding: Druckhaus Beltz, Hemsbach/Bergstr.
2146/3140-543210

Contents

Introduction

There are many fields of stochastics where multi-parameter processes can be encountered. For example, to register the positions of the spins of a ferromagnetic substance at a fixed instant of time, one has to attach an appropriate state space to every point of a three-dimensional lattice. Mathematically this leads to a family of random variables indexed by a subset of $\mathbb{R}^3$, a special case of a so-called stochastic field. Correspondingly, formalizing "multivariate observations" may lead to a stochastic process indexed by a set which, according to its order properties, can be interpreted as a multi-time. The infinite dimensional Ornstein - Uhlenbeck process which appears in a variant of Malliavin's variational calculus, may be considered as a stochastic process with a multi-time or a kind of mixed space-time parameter set (see Ikeda, Watanabe [24]). A close relative of it is the "Wiener sheet" which is undoubtedly the most frequently studied among all multi-parameter processes with a continuous parameter set (see Föllmer [21]). Walsh [43] encounters the Wiener sheet in the study of mathematical models which may arise in neurophysiology or also in problems related to heat conduction or electrical cables. We finish our selection with a more recent example. The investigation of the "Poisson chaos" seems to produce a new kind of infinite dimensional Ornstein - Uhlenbeck process which, considered as a stochastic process indexed by a two-dimensional continuous variable, behaves like a Poisson process in one direction and like a Brownian motion or a more general Gaussian process in the other direction (see Ruiz de Chavez [39], Surgailis [41], [42]).

This book is meant as a contribution to the foundations of a general theory of multi-parameter processes and their stochastic calculus. Like most of the authors who have studied this theory since 1975, when a first pioneering paper of Cairoli, Walsh [12] appeared, we restrict our attention to two-parameter processes. We have a good reason to do so, which might be underestimated at present but will become clear along the way. The considerable degree of complication we will have to face may be compensated by the geometrical simplicity of the notions and results in a two-parameter setting. On the other hand, we also have good reasons to hope that our results can be extended to an arbitrary finite number of parameters.

In the one- and two-parameter theory alike, martingales - the processes we will investigate - play an equally important and central role. An integrable process M which, like all two-parameter processes considered here, is a family of random variables on our basic probability space $(\Omega,\mathcal{F},P)$ indexed by $[0,1]^2$, is called martingale with respect to our basic filtration $(\mathcal{F}_t)_{t\in[0,1]^2}$ (i.e. a family of σ-algebras in $\mathcal{F}$, where $\mathcal{F}_t$ represents the information available at $t\in[0,1]^2$, which is increasing with respect to coordinatewise linear order on $[0,1]^2$) if M_t is $\mathcal{F}_t$-measurable (i.e. M is adapted) and for $s \leq t$ conditioning M_t by $\mathcal{F}_s$ gives M_s. One of the primary aims of an advanced martingale theory consists in the development of a "stochastic calculus" which is the basis for the field of stochastic differential or integral equations and the stochastic counterpart of the classical infinitesimal calculus. Its main theorem, known as "Itô's formula", corresponds to the fundamental theorem of calculus relating differentiation and integration. Given a function x of bounded variation on

$[0,1]^2$ which vanishes on the axes, and a C^1-function f, the classical fundamental theorem states that $f(x_{(1,1)})-f(0)$ is given by the integral of $f'(x)$ with respect to the measure dx defined by the variation of x over $[0,1]^2$. Correspondingly, the simplest version of Itô's formula is for processes X on $\Omega\times[0,1]^2$ vanishing on the axes and whose trajectories $X(\omega)$, $\omega\in\Omega$, have bounded variation. It is given by the classical formula, trajectory by trajectory, the random variational measure $dX(\omega)$, $\omega\in\Omega$, replacing dx. As it happens, many interesting stochastic processes, like for example the Brownian motion and its two-parameter analogon, the Wiener sheet, have infinite variation, i.e. their trajectories are non-rectifiable curves. One of the main achievements of early stochastic calculus for one-parameter martingales M was to realize that in its fundamental theorem besides the "variational integral" with respect to dM, which corresponds to the classical one and becomes now a "stochastic integral", a second order term appears. It is an integral of the second derivative of f with respect to the "quadratic variation" of M. Given a sequence of partitions of the interval [0,t] by intervals whose mesh goes to zero, the quadratic variation $[M]_t$ of M at t can be defined as the limit in probability of the sequence of sums of squared increments of M over the intervals of a partition. Taking a key position in Itô's formula, the quadratic variation process is one of the most important basic processes of stochastic analysis. This is equally true for the theory of two-parameter processes, for which quadratic variation is analogously defined with respect to intervals in $[0,1]^2$.

However, the problem of the existence of quadratic variation for two-parameter martingales proved to be tough, and the progress in solving it slow. In 1981, Zakai [46] established existence for

L^4-integrable continuous martingales, extending the proof of Cairoli, Walsh [13] for martingales of the Wiener sheet. In 1984, Nualart [36] succeeded in generalizing this result to square integrable continuous martingales. For more general (especially non-continuous) martingales, only few fragmentary results were available (see Frangos, Imkeller [22] and the references there). One One of the two main subjects of this book is to derive the existence of quadratic variation for square integrable and, more generally, $L \log^+ L$-integrable two-parameter martingales. The second one lies in the method applied to accomplish this aim. It consists in deriving a representation theorem for square integrable martingales by various "pure jump parts" and a "continuous part" and constructing their quadratic variation from those of the parts. This theorem is the end-product of a procedure of "reduction of jumps" which stimulates a deeper study of the structure of martingales and their general theory and is of independent interest, as well. The execution of this entire program, however, attains a degree of complication which makes us prefer to shape its contours at first in the case of a one-parameter martingale M indexed by [0,1] with respect to a filtration $(\mathbb{G}_t)_{t\in[0,1]}$. This will serve as a "red thread" later.

The significance of the concept of "stopping" is one of the big differences between one- and two-parameter theory. Whereas in the former "stopping times" take an eminent place and influence almost all methods of investigation, "stopping points", "stopping lines" and related notions are more peripheric and of limited use in the latter theory. The procedure of separation of jump parts and computation of quadratic variation we are about to sketch, has to take this into account. It will not make any essential use of a stopping notion. Therefore, even in the classical one-parameter martingale theory, although the results are known, the notions and methods we use to derive them seem to bear some novelty.

As one of the most striking elementary phenomena about M, the expected number of up- and downcrossings of a given space interval is bounded. This readily leads to one of the important regularity results of the theory: M possesses a version whose sample paths are right-continuous and have left limits. We denote this regular version again by M and can now talk about "jumps". A point (ω,t) in $\Omega\times[0,1]$ is called jump of M, if $M_t(\omega)-M_{t-}(\omega)\neq 0$, where $M_{t-}(\omega)$ is the left hand limit of $M(\omega)$ at t. Now the observation made above can be put in more stringent terms: for any $n\in\mathbb{N}$, the random set S_n of jumps of M of heights between $\frac{1}{n}$ and $\frac{1}{n-1}$ has $[0,1]$-sections whose cardinalities constitute an integrable random variable. In particular, they are finite a.s..

Assume that M is square integrable in addition. We plan to get a decomposition of M by a jump part and a continuous part (both square integrable martingales) by successively extracting the jumps of M on S_n, $n\in\mathbb{N}$. Fix $n\in\mathbb{N}$ and consider the jump process M_n of M, restricted to S_n, which, at time t, is just the sum of all jumps of M on S_n up to t. The process M_n is of bounded variation, but, of course, need not be a martingale. Yet, the part we want to cut off M in order to obtain a continuous remainder, is to be a martingale. Therefore, our task could be put in the following terms: compensate M_n by a process C_n of integrable variation in such a way that firstly the resulting process M_n-C_n is a martingale and secondly C_n creates no new jumps, i.e. C_n is continuous. The first requirement will ensure that the decomposition we obtain is a martingale decomposition, the second that the remainder after finishing the cut-off procedure is indeed continuous.

The solution of this compensation problem requires a closer study of measurability concepts in the product $\Omega\times[0,1]$. To see which σ-algebra might play a role, let us try to find conditions under which we face the simplest case - the case in which M_n is already a martingale and compensation unnecessary. First assume that $S_n = \Omega\times\{t_o\}$ is a deterministic set, for some $t_o\in[0,1]$. Let $(t_m)_{m\in\mathbb{N}}$ be a strictly increasing sequence in parameter space which converges to t_o, and let $X^m = (M_{t_o}-M_{t_m})\, 1_{[t_o,1]}$. Now the martingale property of M states that for any pair (u,v) of times, $u\leq v$, conditioning M_v by $\mathfrak{G}_u$ gives M_u. This implies that X^m_v, conditioned by $\mathfrak{G}_u$, may only differ from X^m_u if $t_m<u<t_o$. But, as becomes evident from Doob's maximal inequality, the convergence $X^m \to M_n$ as $m\to\infty$ is dominated and thus M_n turns out to be a martingale. Next assume that S_n is a random one point set. It can easily be seen to be just the graph of a stopping time T_o with respect to $(\mathfrak{G}_t)_{t\in[0,1]}$. Since the conditioning property characterizing a martingale extends from a pair of deterministic times (u,v) to a pair of stopping times (U,V), $U\leq V$, we see that M_n is a martingale, if, as above, T_o can be "predicted" by a strictly increasing sequence $(T_m)_{m\in\mathbb{N}}$ of stopping times. Of course, this need not necessarily be the case. It is true if and only if T_o is measurable with respect to the σ-algebra of "previsible sets" which is generated by all continuous adapted processes. "Continuity" ensures that stopping times in this σ-algebra can be predicted. There is another important σ-field in $\Omega\times[0,1]$, in which this is typically not the case. It is generated by the regular adapted processes, hence contains the previsible sets and is called σ-algebra of "optional sets". Of course, M and therefore S_n is optional. And, as is suggested by what has been said above, M_n

is a martingale if S_n is previsible.

Let us return to the problem of compensation of M_n. A version of the decomposition theorem of Doob-Meyer points in the direction of a possible solution. This theorem which we will refer to as "projection theorem" is of central importance for our analysis and states that for any integrable increasing process A there exists a unique previsible increasing process A^p, called its "dual previsible projection" such that $A-A^p$ is a martingale. We may apply it to the increasing processes $\overline{M}_n$, $\underline{M}_n$ of positive resp. negative jumps of M on S_n. Then $C_n = \overline{M}_n^p - \underline{M}_n^p$ makes $M_n - C_n$ a martingale. But to be a good candidate for a compensator, C_n has to be continuous, too. This need not be true in general. Yet, the above discussion already indicates what might be crucial for the problem. According to it, the previsible case is already solved. Assume now, on the contrary, that S_n intersects any previsible random set of the same geometric type (i.e. its [0,1]-sections are a.s. finite) only on a negligible set. In this case, S_n is said to be "totally inaccessible", and C_n has to be continuous for the following reasons. In consequence of its previsibility and regularity, it is unable to realize the totally inaccessible set S_n and therefore cannot jump on it. On the other hand, it cannot jump outside S_n, since it can be seen without effort that a previsible martingale of bounded variation, like a hypothetical jump of $M_n - C_n$ on a previsible set outside S_n, vanishes. This important observation indicates how the problem can be attacked in general: partition the optional set of jumps of M by random sets which are as simple as S_n, but which are either previsible or totally inaccessible. Of course, this partition need not coincide with $(S_n)_{n \in \mathbb{N}}$.

We first state more precisely what we mean by "simplicity". Call an optional random set S in $\Omega \times [0,1]$ simple, if the cardinalities of its [0,1]-sections define an integrable random variable. The crux of the construction of the desired partition is then to see how a given simple set (like S_n) can be covered by a sequence of pairwise disjoint either previsible or totally inaccessible sets. To this end, associate with a given simple set S the integrable increasing process $\Gamma(S)$ which counts the points of S, i.e. $\Gamma(S)(\omega)$ at time t is just the cardinality of the intersection of S_ω and $[0,t]$, $\omega\in\Omega$. Consider its dual previsible projection $\Gamma(S)^p$. Since this process is previsible, its jumps can be arranged on a countable union of pairwise disjoint simple random sets $(T_n)_{n\in\mathbb{N}}$ which are previsible. Since $\Gamma(S)-\Gamma(S)^p$ is a martingale, the union of the T_n is already the essential supremum of previsible simple sets in S and its complement in S totally inaccessible. Hence, an analysis of the jumps of the dual previsible projection of the increasing process associated with a simple set gives the desired covering sequence.

Assume now a partition $(U_n)_{n\in\mathbb{N}}$ of the set of jumps of M by either previsible or totally inaccessible simple sets is given. Let M_n denote again the jump process of M on U_n, C_n its compensator. We know C_n is a continuous process of bounded variation. In the Hilbert space of square integrable martingales whose norm is defined by the usual L^2-norm, the martingales M_n-C_n prove to be pairwise orthogonal due to the pairwise disjointness of the U_n. The orthogonal complement of the sum M^o of these compensated jumps defines a square integrable continuous martingale M^c, due to the continuity of the compensators. This finishes the procedure of decomposing M by a pure jump martingale M^o and a continuous martingale M^c.

Given this decomposition, it is not hard to compute the quadratic variation of M. One simply has to compute the quadratic variations of M^o and M^c and add them: the fact that all compensators are continuous and of bounded variation, hence cannot contribute, and that M^c is continuous, makes the quadratic variations of the orthogonal parts also "orthogonal". Now the quadratic variation of M^o is just the sum of the squares of the jumps of M. To obtain the quadratic variation of M^c, one may take resort to a more general version of the theorem of Doob-Meyer than the above projection theorem. It states that for any nonnegative submartingale X there exists a unique previsible increasing process A, also called "dual previsible projection" of X, such that X-A is a martingale. It applies to $(M^c)^2$ and allows to identify the dual previsible projection of this process as the quadratic variation of M^c.

To summarize, the procedure of investigating the structure of square integrable martingales M and their quadratic variations which has been outlined consists of the following main steps:

1) by application of an appropriate version of the decomposition theorem of Doob-Meyer (called projection theorem in the case of increasing processes) find the dual previsible projections of increasing processes (like $\Gamma(S)$ for simple S) and of submartingales (like M^2),
2) partition the random set of jumps of M by a sequence $(U_n)_{n \in \mathbb{N}}$ of either previsible or totally inaccessible simple sets,
3) find the compensators C_n of the jump process M_n of M on U_n and show they are continuous,
4) show that the compensated jump processes M_n-C_n are pairwise orthogonal, subtract their sum M^o from M to obtain the continuous part,
5) compute the quadratic variations of M^o and M^c and sum up to obtain the quadratic variation of M.

Guided by the above 5 step program we will now give an outline of an analogous procedure of gradual elimination of jumps and computation of quadratic variation for two-parameter martingales, as presented in this book. Of course, the notions and statements figuring in 1)-5) have to be reinterpreted in a two-parameter setting. As a usual phenomenon which might be realized rather soon, the theory hereby becomes essentially more complicated. For this reason, we emphasize that our "sketch" can be considered as being of a more than purely introductory character. On one hand, we felt that at places it might become too tough to read without offering the reader the opportunity to jump to the details in the text. We therefore have provided it with hints indicating the numbers of corresponding main theorems (T) or propositions (P) of the text. On the other hand, given that the procedure proposed is not of a straightforward kind, this outline might be helpful for a better understanding. It strictly follows the strain of ideas involved in the step-by-step order of 1)-5), not necessarily the order in which the results are presented in the text and which is imposed by logical or formal aspects. Whenever the reader looks for a "red thread" he may jump back to this sketch to regain orientation.

Our analysis starts with a crucial assumption stated first by Cairoli, Walsh [12] in their pioneering paper and adopted ever since by most of the authors concerned with this theory. It says that the information gained in the future of a given time point t in one direction is independent of the information gained in the other direction, given what happened up to t. This "conditional independence" assumption is vital for what follows. It finds its most useful interpretation for our purposes in terms of "optional and previsible (dual) projections". To explain these notions, let us briefly return to the one-parameter case, where we already encountered the dual previsible projection of an integrable increasing process. The theorem stating its existence can be viewed in an alternative way. For a given integrable increasing process A, let m_A denote the measure on the product $\Omega\times[0,1]$ defined by $m_A(S) = E(\int 1_S\, dA)$, S a product measurable set. Then it states that there is exactly one previsible increasing process A^p such that the integrals of any bounded previsible process with respect to m_A and m_{A^p} are the same. An optional version of this theorem says that there is exactly one optional increasing process A^o (the "dual optional projection" of A) such that the integrals of any bounded optional process with respect to m_A and m_{A^o} coincide. Finally, the statements of the projection theorems can be dualized. Given a bounded product measurable process X, there is exactly one bounded optional (previsible) process ${}^oX$ (${}^pX$) such that X and ${}^oX$ (X and ${}^pX$) cannot be distinguished by measures associated with optional (previsible) integrable increasing processes. ${}^oX$ resp. ${}^pX$ is called "optional" resp. "previsible" projection of X. Projection theorems and the martingale notion are linked by the following elementary observation. If X is a bounded random variable, considered as a process with a trivial time dependence, and M is a regular version of the martingale generated by taking conditional expectations of X, then ${}^oX=M$, ${}^pX=M_-$, where M_- is the process of left limits of M. This identification is the starting point for our study of projections for two-parameter processes.

We first prove that "one-directional" (dual) projections exist

(T(4.1), T(4.2)). More precisely, for a given product measurable bounded process X on $\Omega\times[0,1]^2$ the family of optional (previsible) projections of the one-parameter processes $X_{(\cdot,r)}$ in i-direction, r being the (fixed) parameter of the complementary direction, can be chosen measurably with respect to the product $\mathfrak{G}^i$ ($\mathfrak{P}^i$) of the optional (previsible) sets in i-direction with the Borel sets in the remaining one, called σ-algebra of "i-optional" ("i-previsible") sets, i=1,2. A dual statement can be made for integrable increasing processes A, where "increasing" means that they define random measures on the Borel sets of $[0,1]^2$ in the usual way. The resulting processes ${}^{\gamma_i}X$ (${}^{\pi_i}X$) resp. A^{γ_i} (A^{π_i}) are called (dual) i-optional (i-previsible) projections. A series of regularity results (T(6.1), T(6.2)) shows that X inherits regularity properties to its projections ${}^{\gamma_i}X$ (${}^{\pi_i}X$). They in turn imply that any $L\log^+L$-integrable martingale M possesses a version whose trajectories are continuous for approach in the right upper quadrant and possess limits for approach in the left upper and right lower quadrants (P(8.2)). Strangely enough, only an additional investigation of a stochastic integral for square integrable martingales which shows that they can be "stopped" on more complicated random sets (P(9.3), P(9.4)) reveals that M also possesses limits for the left lower quadrant (P(9.5)). To sum up, $L\log^+L$-integrable martingales have "regular" versions (T(9.1)). Once this is established, it is not hard to link two-parameter projections and martingales in the same way as above. If X is a bounded random variable, M a regular version of the martingale generated by taking conditional expectations of X, then ${}^{\gamma_1\gamma_2}X = M = {}^{\gamma_2\gamma_1}X$, ${}^{\pi_1\gamma_2}X = M^{-\cdot} = {}^{\gamma_2\pi_1}X$, ${}^{\gamma_1\pi_2}X = M^{\cdot-} = {}^{\pi_2\gamma_1}X$, ${}^{\pi_1\pi_2}X = M^{--} = {}^{\pi_2\pi_1}X$ (T(10.1)). Here "-" always means that left limits have to be taken in the respective direction, for example: $M^{-\cdot}$ is the process defined by the left limits of M in the first direction. This result has very important consequences, once it is extended to general bounded product measurable processes and dualized. It says that it is immaterial in which order iterated (dual) projections are taken and allows us to define (dual) previsible (i.e. 1- and 2-previsible, denoted by ${}^{\pi}X$ resp. A^{π}) projections, 1-previsible, 2-optional projections, 2-previsible, 1-optional projections, and optional (i.e. 1- and 2-optional, denoted by ${}^{Y}X$, A^{Y}) projections by just the product of two one-directional projections in an arbitrary order (P(10.1)). Here is what we mean by the most useful interpretation of the conditional independence assumption: projections and dual projections in different directions are "independent" of each other, commute and their compositions define "two-directional" (dual) projections in an unambiguous way (T(10.2), T(10.3)). Implicitly we also have accomplished the first half of step 1 of our program. Given an integrable increasing process A, we know that $A - A^{\pi_1}$ is a martingale in direction 1, whose dual previsible projection is $A^{\pi_2} - A^{\pi_1\pi_2} = A^{\pi_2} - A^{\pi}$. Hence $A - (A^{\pi_1} + A^{\pi_2} - A^{\pi})$ is a martingale in both directions, i.e. a martingale (cor. 1 of T (11.2)). At this stage, the second half of step 1 is not very difficult any more. Call an integrable process X submartingale, if it is a submartingale in the order sense, i.e. if it is a one-parameter submartingale on every increasing path in $[0,1]^2$, and a weak submartingale, if the increments $\Delta_J X$ of X over rectangles J in parameter space, conditioned by the information available at their lower left boundary points, are nonnegative. We concentrate on the construction of the counterpart of A^{π} for X. The process A^{π} is the uniquely determined previsible increasing process such that $m_A|\mathcal{P} = m_{A^{\pi}}|\mathcal{P}$, where $\mathcal{P} = \mathcal{P}^1 \cap \mathcal{P}^2$ is the σ-algebra of previsible sets and m_B defined for integrable increasing B like in the one-parameter case. For previsible rectangles $F\times J$ (i.e. $J=]s,t]$ interval in $[0,1]^2$, $F\in\mathcal{F}_s$), set $m_X(F\times J) = E(1_F\, \Delta_J X)$. If X is a weak submartingale, m_X is a nonnegative set function. Since the previsible rectangles generate $\mathcal{P}$, according to what has been said about A, the existence of a "dual previsible projection" of X is proved, once we know that m_X has a σ-additive extension to $\mathcal{P}$, called "Doléans measure" of X. In the theory of one-parameter processes, the existence of Doléans measures for submartingales has been characterized by a criterion which states the uniform integrability of the family of all random variables generated by stopping X at stopping times taking only finitely many values. In lack of a good notion of stopping for our theory, we characterize it by a related uniform integrability condition involving approximations of the dual previsible projections of X (T(5.1), T(11.1)). We thus obtain the following result. If X is a submartingale and a weak submartingale which fulfills this condition together with its "marginal" processes $X_{(.,1)}$, $X_{(1,.)}$, it is decomposed by a martingale, two processes which are martingales in one direction and of bounded variation in the other one, and a previsible increasing process (T(11.2) and the preparatory T(5.2)). This decomposition theorem applies in particular to the square of a square integrable martingale (cor.2 of T(11.2)).

To turn to steps 2)-5), we first have to know where to look for discontinuities of regular two-parameter martingales M. The optional random set of points in $\Omega\times[0,1]^2$ where M has discontinuities is contained in a random countable union of vertical and horizontal line segments in $[0,1]^2$ (T(1.1)). But here is an instance of complication. We cannot simply call the discontinuities of M "jumps". Given a regular real-valued function f, one can distinguish three kinds of discontinuities: a point t is called

"0-jump of f", if $\Delta_t f = f(t) - f^{-\cdot}(t) - f^{\cdot -}(t) + f^{--}(t) \neq 0$, "1-jump of f", if $f(t) - f^{-\cdot}(t) \neq 0$ and $\Delta_t f=0$, "2-jump of f", if $f(t) - f^{\cdot -}(t) \neq 0$ and $\Delta_t f=0$. We will have to deal with all three kinds of jumps separately.

Assume M is a square integrable regular martingale. We propose to decompose M into various jump martingale parts and a continuous martingale part in the following way. We first treat the 0-jumps of M exclusively and run through a procedure along the lines of 2)-4) to separate a 0-jump part. Its orthogonal complement with respect to M in the Hilbert space of square integrable martingales will turn out to be free of 0-jumps and possess at most 1- and 2-jumps. Next we run through the same procedure two more times to separate a 1-jump part and a 2-jump part of M. The orthogonal complement of the three jump parts will be free of jumps of any kind, i.e. continuous. This will yield the desired orthogonal decomposition of M. Finally, we will consider the problem of the existence of quadratic variation, extending the ideas of 5).

We turn to the 0-jumps of M to execute step 2). For $n \in \mathbb{N}$, the optional set of points in $\Omega \times [0,1]^2$ where M has 0-jumps of heights between $\frac{1}{n}$ and $\frac{1}{n-1}$ has $[0,1]^2$-sections whose cardinalities constitute an integrable random variable (T(14.1)). This indicates what simple sets in the context of 0-jumps should be. Call an optional random set "0-simple", if the cardinalities of its $[0,1]^2$-sections define an integrable random variable. The essence of step 2) for 0-jumps (T(14.1)) now consists in considering a 0-simple set L and finding a covering of L by a countable family of pairwise disjoint previsible or inaccessible 0-simple sets. Here is another instance of complication. We have to distinguish four different types of inaccessibility and previsibility. They will emerge in the following discussion in which we analyze the dual previsible projections of the increasing process $\Gamma(L)$ associated with L, as is suggested by the one-parameter case. For $t\in[0,1]^2$, $\Gamma(L)_t$ is the number of points in $L\cap[0,t]$. The 0-jumps of the dual previsible projections $\Gamma(L)^{\pi_i}$ can be arranged on two sequences of pairwise disjoint 0-simple sets $(U_n^i)_{n\in\mathbb{N}}$, i=1,2. We subtract from L the union of these sequences and obtain a "totally inaccessible" remainder T. The sets U_n^i are i-previsible, since $\Gamma(L)^{\pi_i}$ is, but they still have to be treated in the complementary direction. So we repeat the procedure just executed for any of these sets. Consider the dual previsible projection $\Gamma(L)^{\pi} = \Gamma(L)^{\pi_1\pi_2} = \Gamma(L)^{\pi_2\pi_1}$ of $\Gamma(L)$. Arrange its 0-jumps on a countable family of pairwise disjoint 0-simple sets $(T_m)_{m\in\mathbb{N}}$. Subtract the union of the T_m from each U_n^i, $n\in\mathbb{N}$, i=1,2. This leaves two sequences $(T_n^i)_{n\in\mathbb{N}}$ of "1-previsible, 2-inaccessible" resp. "2-previsible, 1-inaccessible" 0-simple sets. They do not overlap in consequence of the commutability of π_1 and π_2. To sum up, L is contained in the union of the pairwise disjoint 0-simple sets T, T_n^i, T_n, $n\in\mathbb{N}$, which are totally inaccessible resp. i-previsible, (3-i)-inaccessible resp. previsible, i=1,2 (T(13.1)).

For step 3), we have to deal with the compensation of jumps on 0-simple sets. Assume L is a 0-simple set of one of the four types just discussed. The jump process M(L) of M on L at t is simply the sum of all 0-jumps of M on L up to t. In the simplest case, L is previsible (i.e. 1- and 2-previsible). Like in the one-parameter case, M(L) turns out to be a martingale in both directions, hence a martingale and need not be compensated (P(15.1)). If L is 1-previsible, 2-inaccessible, we take $C^2 = \overline{M}(L)^{\pi_2} - \underline{M}(L)^{\pi_2}$, where $\overline{M}(L)$, $\underline{M}(L)$ are the positive resp. negative jumps of M on L. Then C^2 is still a martingale in direction 1, since M(L) is. But it is 2-previsible and therefore, L being 2-inaccessible, can have at most 1-jumps. In particular, it is free of 0-jumps. Since $M(L)-C^2$ is a martingale, too, C^2 is appropriate as a compensator - in a wider sense, however, than in step 3) above. Although not necessarily continuous, it has essentially better continuity properties than the process it compensates (T(15.2)). In the most difficult case, L is totally inaccessible. Let $C^i = \overline{M}(L)^{\pi_i} - \underline{M}(L)^{\pi_i}$, $C = \overline{M}(L)^{\pi} - \underline{M}(L)^{\pi}$. Then C^i has at most (3-i)-jumps and C is continuous. The fact that π_1 and π_2 commute makes $M(L)-(C^1+C^2-C)$ a martingale. Therefore C^1+C^2-C is appropriate as a compensator in the above mentioned wider sense (T(15.1)). To summarize, any jump process on a 0-simple set of one of the four inaccessibility degrees possesses a compensator free of 0-jumps.

It is not hard to prove that compensated jump parts belonging to disjoint 0-simple sets are orthogonal (P(8.3), P(18.1)). Moreover, since 1- and 2-jumps are essentially different from 0-jumps, a compensated 0-jump is orthogonal to any martingale possessing at most 1- or 2-jumps (P(18.2)). The continuity properties of the compensators imply that the subtraction of the orthogonal sum M^o of compensated 0-jumps from M produces a square integrable martingale which possesses at most 1- and 2-jumps and is orthogonal to M^o(P(19.1), T(19.1)). This finishes step 4) for 0-jumps.

To start again with step 2) for 1- jumps, assume now that M is a square integrable martingale without 0-jumps. For $n \in \mathbb{N}$, the optional set of points in $\Omega\times[0,1]^2$ where M has 1-jumps of height at least $\frac{1}{n}$, is contained in a random family of vertical line segments in $[0,1]^2$ whose upper boundaries are on $\partial[0,1]^2$, which do not contain their lower boundaries and whose number is integrable (T(14.2)). The class S^1 of optional sets with these properties, however, is not stable with respect to relative complements, unlike their 0-simple counterparts. But, to be able to partition all 1-jumps of M, we need to take relative complements. Therefore we call all sets in the semiring generated by S^1 "1-simple" and define "2-simple" sets in an analogous manner. Since the line segments constituting them do not contain their lower boundary points, 1-simple sets are seen to be 2-previsible (cor. of P(12.1)). As a consequence, the process of isolating inaccessible parts is somewhat less involved for this kind of simple sets. Again, the essence of step 2) for 1-simple sets (T(14.2)) consists in finding a covering of a given 1-simple set L by a sequence of pairwise disjoint either previsible or inaccessible sets of the same kind. We first define an integrable increasing process $\Gamma(L)$ associated with L by taking $\Gamma(L)_t$ to be the total length of the intersections of the line segments of L and $[0,t]$, $t\in[0,1]^2$. By decomposing the 1-jumps of $\Gamma(L)^{\pi_1}$ we find a sequence $(T_n)_{n\in\mathbb{N}}$ of pairwise disjoint previsible sets ($\Gamma(L)^{\pi_1}$ is previsible), whose complement T in L is 1-simple and "inaccessible" (T(13.2)). Of course, for 1-simple sets, inaccessibility is essentially "1-inaccessibility".

To execute step 3) for 1-simple sets, we first have to see what the jump process M(L) of M on a 1-simple set L is. It is essentially different from the jump processes encountered so far. For any $t_1\in[0,1]$, the one-parameter process $M_{(t_1,.)}-\dot{M}^{-}_{(t_1,.)}$ is a martingale which is continuous since M has no 0-jumps. We can define M(L) at t to be the random sum of all these one-parameter martingales along the intersections of the line segments belonging to L and $[0,t]$, $t\in[0,1]^2$. M(L) is a continuous martingale in direction 2 which is of bounded variation in direction 1 by definition of the 1-simple sets (P(14.2)). If L is previsible, for the same reasons as above, M(L) is already a martingale and compensation unnecessary (P(15.2)). If L is inaccessible, a compensator C can be defined by applying a decomposition theorem to the submartingales $\overline{M}(L)$, $\underline{M}(L)$ given by the sum of the positive resp. negative parts of the martingales $M_{(t_1,.)}-M^{-.}_{(t_1,.)}$ along the line segments of L. Inaccessibility of L forbids C to have 1-jumps, continuity of M(L) in direction 2 to have 2-jumps. Hence C proves to be continuous (T(15.3)). For 2-simple sets, an analogous procedure is possible. To summarize, the jump processes of M on 1- or 2-simple sets possess compensators which are again "essentially more continuous" than the processes they compensate.

Turning to step 4), we observe that disjointness of jump sets and essential difference of jump types makes the following martingales orthogonal: two compensated 1-jumps or 2-jumps belonging to disjoint simple sets (P(8.3), P(18.3)), a compensated 1-jump and a compensated 2-jump (P(8.3), P(18.3)), a compensated 1- or 2-jump and a continuous square integrable martingale (P(8.3), P(18.4)). The orthogonal sum of compensated i-jumps of M therefore uniquely defines an i-jump part M^i, i=1,2, such that M^1 and M^2 are orthogonal (P(19.2), T(19.1)). Moreover, the continuity of the compensators ensures that the orthogonal complement M^c of M^1+M^2 is continuous.

To complete the results of the preceding discussion, a general square integrable martingale M possesses pairwise orthogonal i-jump parts M^i, i=0,1,2, and a continuous part M^c which decompose M. The jump parts can be represented as orthogonal sums of the compensated jumps of the respective kind as described above (P(19.3), T(19.1)). This finishes step 4), and we are ready to apply our knowledge to prove the existence and study the structure of the quadratic variation of M (step 5)).

The first observation one can make in this context concerns the compensators of given sequences of jump processes of M. Due

to their variation order - for 0-jumps they are of bounded variation, for i-jumps of bounded variation at least in direction i, i=1,2 - they cannot contribute to the quadratic variation (P(16.2), P(16.4)). One consequence of this and pairwise disjointness of given sequences of simple sets covering the jumps of M is that the quadratic variations of the four components of M turn out to be "orthogonal" and simply sum up to give the quadratic variation of M. As another consequence, they are as expected:
for M^o, the sum of all squares of 0-jumps of M (T(17.1), T(20.1)),
for M^1, the sum of all (continuous) quadratic variations of the one-parameter martingales $[(M-M^o)_{(t_1,\cdot)}-(M-M^o)^{-\cdot}_{(t_1,\cdot)}]$, $t_1\in[0,1]$ (T(17.2), T(20.1)),
for M^2, an analogous result with the axes switched,
for M^c, the continuous dual previsible projection of its square (T(17.3), T(20.1)).
This identification and the fact that continuity properties of compensated jumps on simple sets are easy to handle also gives us complete information about the relationship between continuity properties of M and its quadratic variation [M]. If M has no 0-jumps, [M] does not; if M has no 0-jumps and no i-jumps, [M] has the same property, i=1,2; if M is continuous, so is [M] (T(20.2)).

We finally propose to determine the largest class of martingales which possess a quadratic variation. By approximation via bounded martingales and with the support of a "weak type" of Burkholder's inequality we prove that any $L \log^+ L$-integrable martingale does (T(20.3)). Moreover, we construct an L^1-integrable martingale which possesses no quadratic variation (example in section 21). Hereby, like in many parts of the paper, martingale inequalities of the different types, like Burkholder's (T(8.2), T(21.1)), Doob-Cairoli's and extensions thereof (T(4.4), T(8.1), T(10.5)), or Garsia-Neveu's (T(4.3), T(5.3), T(10.4), T(11.3)) play an eminent role.

The formal organization of the book is as follows. Chapter I covers the "one-directional" part of step 1). More precisely, we prove one-directional projection and decomposition theorems. The basic result of chapter II is the regularity theorem for $L \log^+ L$-integrable martingales, resting upon the conditional independence assumption and implying the relative simplicity of the two-directional projection and decomposition theorems. This is the "two-directional" part of step 1). Chapter III is devoted to steps 2) and 3) of our program. The main tool is the study of simple sets via increasing processes and their dual projections. In chapter IV, finally, steps 4) and 5) are executed. It presents the main results of our investigations.

As is indicated in the introductory remarks of sections 1-11, many results and methods of chapters I and II have been taken from original papers of different authors. However, writing these two chapters took more effort than the rest of the book. One of the main difficulties we had to overcome was to find a unified treatment for the different results available. Except for the short outlines of Meyer [33] and Merzbach [28] this had not been done before. The search for the reasons leads us to a typical problem of the theory of multi-parameter processes. For many questions it offers a variety of approaches, some of which suffer from the fact that an analogous problem has been solved for one-parameter processes and the methods used there are taken over. If they are too closely related to the particularities of a one-parameter setting, they are likely to obscure a genuinely two-parameter treatment. Unfortunately, it is often not easy to decide which approaches are of this kind. This dilemma might still be felt for example in the proof of the regularity theorem for martingales in section 9. The existence of limits in the left lower quadrant is proved by a method using "stopping lines", whereas for the remaining three quadrants "stopping points" are employed. We conjecture that "cracks" like this in the theory can be removed by designing a uniform method which is intrinsically two-parameter.

Acknowledgements

I would like to express my gratitude to Hans-Otto Georgii, Hans Kellerer and Michel Métivier for their support and many helpful comments during the preparation of this book. I also feel much indepted to Louis Sucheston for rendering possible a stay at the Ohio State University in 1984/85 which stimulated the underlying research.

0. Notations and conventions

Unless otherwise explicitly stated, the parameter space of the stochastic processes we consider is $\mathbb{I} = [0,1]^2$. It is coordinatewise linearly ordered by "$\leq$". Intervals with respect to this order are defined in the usual way. For example, $[s,t[$ is the set of all $u \in \mathbb{I}$ such that $s \leq u$, $u < t$, both coordinatewise for $s, t \in \mathbb{I}$, $s \leq t$. We make one exception to this rule: for the set of left open, right closed intervals, which we denote by $\mathcal{J}$, "openness" is always understood with respect to the relative topology of $\mathbb{I}$. For example, $]0,t] = [0,t]$ for $t \in \mathbb{I}$. If J is an interval, we write s^J, t^J for the points in $\mathbb{I}$ which determine J. These conventions are also adopted for intervals in $[0,1]$. By a partition of $\mathbb{I}$ ($[0,1]$) we always mean a partition of $\mathbb{I}$ ($[0,1]$) generated by a finite number of axial parallel lines (points) in $\mathbb{I}$ ($[0,1]$) consisting of left open, right closed intervals. A 0-sequence of partitions of $\mathbb{I}$ ($[0,1]$) is a sequence of partitions of $\mathbb{I}$ ($[0,1]$) which is increasing with respect to fineness and whose mesh converges to 0. To denote components of time points in $\mathbb{I}$ of intervals or partitions we use a lower subscript. For example, $t = (t_1,t_2) \in \mathbb{I}$, $J = J_1 \times J_2$ or $\mathbb{J} = \mathbb{J}_1 \times \mathbb{J}_2$ for an interval J or a partition $\mathbb{J}$. If one of these lower indices is i, the complementary index $3-i$ is simply written $\bar{i}$. Given an interval $J \in \mathcal{J}$, $i \in \{1,2\}$, the "i-shadow" of J is the interval $J^i =](0,s_{\bar{i}}^J), (s_{\bar{i}}^J, t_i^J)]$ (we sometimes write $t = (t_i, t_{\bar{i}})$ regardless of whether $i=1$ or 2 for $t \in \mathbb{I}$).

The symbol $|A|$ is used for the cardinality of a set A. It will always be clear from the context what $[a]$ or $[f]$ means:

if $a \in \mathbb{R}$, it is the integer part of a, if f is a function, it denotes the graph of f. Given a function $f : [0,1] \to \mathbb{R}$, the increment of f over an interval J in [0,1] will be denoted by $\Delta_J f$. This also applies to functions $f : \Pi \to \mathbb{R}$. Here

$$\Delta_J f = f(t^J) - f(s_1^J, t_2^J) - f(t_1^J, s_2^J) + f(s^J).$$

f is called increasing, if $\Delta_J f \geq 0$ for all intervals J. The variation of f is $\text{var}(f) = \sup\{ \sum_{J \in \mathbb{J}} |\Delta_J f| : \mathbb{J}$ partition of Π ([0,1])$\}$, and f is said to be of bounded variation, if $\text{var}(f) < \infty$.

If $(X,\mathcal{A})$, $(Y,\mathcal{B})$ are measurable spaces, $\mathfrak{M}(\mathcal{A},\mathcal{B})$ denotes the set of $\mathcal{A}$-$\mathcal{B}$-measurable functions mapping X to Y. The σ-algebra generated by a system $\mathcal{C}$ of sets in $\mathcal{P}(X)$ is $\sigma_X(\mathcal{C})$. For the Borel sets on a subset C of $\mathbb{R}$ or $\mathbb{R}^2$ we use the symbol $\mathcal{B}(C)$, for the Lebesgue measure λ resp. λ^2. For measures μ and ν, $\mu \ll \nu$ expresses that μ is absolutely continuous with respect to ν. Our basic probability space is denoted by $(\Omega,\mathcal{F},P)$. $\mathcal{F}$ is assumed to be complete with respect to P. The filtration $\mathbb{F} = (\mathcal{F}_t)_{t \in \Pi}$ which is always supposed to be given satisfies some basic assumptions: it is right-continuous, i.e. $\mathcal{F}_s = \bigcap_{t>s} \mathcal{F}_t$ for all $s \in \Pi$, it is complete, i.e. every $\mathcal{F}_t$ contains all P-zero-sets and, for convenience, $\mathcal{F}_t$ is trivial for $t \in \Pi \cap \partial \mathbb{R}_+^2$. The most important hypothesis, however, without which our analysis would be completely impossible, is the "conditional independence" of the marginal filtrations $\mathbb{F}_1 = (\mathcal{F}_{t_1}^1)_{t_1 \in [0,1]}$ and $\mathbb{F}_2 = (\mathcal{F}_{t_2}^2)_{t_2 \in [0,1]}$, where $\mathcal{F}_{t_1}^1 = \mathcal{F}_{(t_1,1)}$, $\mathcal{F}_{t_2}^2 = \mathcal{F}_{(1,t_2)}$. It states that $\mathcal{F}_{t_1}^1$ and $\mathcal{F}_{t_2}^2$ are conditionally independent given $\mathcal{F}_t$ for all $t \in \Pi$, or equivalently that for any bounded random variable X we have

$$E(X|\mathcal{F}_t) = E(E(X|\mathcal{F}_{t_1}^1)|\mathcal{F}_{t_2}^2) = E(E(X|\mathcal{F}_{t_2}^2)|\mathcal{F}_{t_1}^1) \text{ for } t \in \Pi.$$

It is supposed to be always fulfilled and particularly implies that $\mathcal{F}_t = \mathcal{F}^1_{t_1} \cap \mathcal{F}^2_{t_2}$ for $t \in \Pi$. The most important measurable spaces we will encounter are connected with the product sets $\Omega \times \Pi$ or $\Omega \times [0,1]$. By π_Ω we denote the projection of $\Omega \times \Pi (\Omega \times [0,1])$ on Ω, by $\pi_{\Omega,i}$ the projection of $\Omega \times \Pi$ on the product of Ω and the ith coordinate interval, $i=1,2$. If $U,V \subset \Omega \times \Pi$, the sets U^a, U^o are defined by $(U^a)_\omega = (U_\omega)^a$, $(U^o)_\omega = (U_\omega)^o$, where C^a resp. C^o is the closure resp. open kernel of a set $C \subset \Pi$. Furthermore, $\text{diam}\,(U)_\omega = \text{diam}\,(U_\omega)$ where $\text{diam}\,(C) = \sup\{|s-t|: s,t \in C\}$ for $C \subset \Pi$ and $U+V = \{(\omega,s+t): s \in U_\omega,\ t \in V_\omega\} \cap \Omega \times \Pi$. The "evanescent sets" in $\Omega \times \Pi$ resp. $\Omega \times [0,1]$, both denoted by the same symbol $\mathbb{N}$, are those sets whose Ω-projections are P-zero. Stochastic processes X are a priori not more than mere families of random variables X_t indexed by $t \in \Pi$. X is called measurable, if $X \in \mathfrak{M}(\mathcal{F} \times \mathcal{B}(\Pi), \mathcal{B}(\mathbb{R}))$. It defines two families of one-parameter processes: for $t_{\bar{i}} \in [0,1]$, $X_{(\cdot,t_{\bar{i}})}$ is the process $(\omega,t_i) \to X_t(\omega)$, $i=1,2$. Random variables or one-parameter processes are sometimes considered as two-parameter processes by just assuming a trivial dependence on t or the second parameter. Two processes are considered as being equal, if they differ on a negligible set. A process X is considered a "version" of a process Y, if for all $t \in \Pi$ we have $X_t = Y_t$ (a.s.). For simplicity, the processes we consider usually are assumed to be zero on $\Pi \cap \partial \mathbb{R}^2_+$. At places, however, this rule is violated (especially in sections 4 and 10 where projections of processes are studied). But it is clear from the context where this might happen. For a process X and a mapping $S: \Omega \to \Pi$, X_S is the mapping $\omega \to X_{S(\omega)}(\omega)$.

At different places it is inevitable that we take over results from the theory of one-parameter processes. For unexplained notations and results in this context the reader is usually referred to Dellacherie, Meyer [17] and [18]. We do not hesitate, however, to explain the most important ones at this place. Whenever a filtration is involved in talking about a one-parameter process and nothing else is said, it is meant to be $\mathbb{F}_1$ or $\mathbb{F}_2$, the context making clear which one. The following notions are with respect to $\mathbb{F}_1$. An $\mathbb{F}_1$-stopping time T is a mapping from Ω to $[0,1]$ such that $\{T\leq r\} \in \mathcal{F}^1_r$ for $r\in [0,1]$. If S,T are $\mathbb{F}_1$-stopping times, $S\leq T$, the stochastic intervals $]S,T]$, $[S,T]$, etc. can be defined. For example, $]S,T[= \{(\omega,r): S(\omega)<r<T(\omega)\}$. With every $\mathbb{F}_1$-stopping time T are associated two σ-algebras: the "past" $\mathcal{F}^1_T = \{A\in\mathcal{F}: A \cap \{T\leq r\} \in\mathcal{F}^1_r \text{ for } r\in [0,1]\}$, and the "strict past" $\mathcal{F}^1_{T-} = \sigma_\Omega(\{r<T\}\cap F: F\in\mathcal{F}^1_r, r\in [0,1])$. If $F\in\mathcal{F}^1_T$, the mapping $T_F = T1_F+1_{\bar{F}}$ is also an $\mathbb{F}_1$-stopping time. Given a square integrable one-parameter martingale M, we denote as usual by $\langle M\rangle$ the dual previsible projection of M^2 and by $[M]$ the quadratic variation of M. The same symbols will be used in chapters II and IV for their two-parameter counterparts.

The most important among the topologies on the subspaces of the space $\mathfrak{M}(\mathcal{F},\mathcal{B}(\mathbb{R}))$ we consider are the usual L^p-topologies which are generated by the functionals $\xi \to E(|\xi|^p)^{1/p}$ resp. $E(|\xi|^p)$ resp. $E(|\xi|\wedge 1)$ for $p\geq 1$ resp. $0<p<1$ resp. $p=0$. They are denoted by $\|\cdot\|_p$ and the corresponding topological vector spaces by $L^p(\Omega,\mathcal{F},P)$. In the convex case, we will have the opportunity to work with more general spaces, the most important among which are generated by "Orlicz" norms belonging to "Young functions" Φ. We only give the main definitions. For more information the reader is referred to Neveu [35] or Dellacherie, Meyer [18], pp. 178-188. Let ϕ be an increasing, right-continuous function on $\mathbb{R}_+$, $\Phi(t) = \int_0^t \phi(s)\,ds$. Then Φ is convex, increasing and $\Phi(0)=0$. Let $p = \sup_{t\in\mathbb{R}_+} \frac{t\phi(t)}{\Phi(t)}$, $\bar{p} = \inf_{t\in\mathbb{R}_+} \frac{t\phi(t)}{\Phi(t)}$, $q = \frac{\bar{p}}{\bar{p}-1}$. Then Φ is called "moderate", if $p<\infty$, p the "exponent" of Φ. If ϕ is unbounded, ψ its right-continuous inverse, the function $\Psi(t) = \int_0^t \psi(s)\,ds$ is well defined, convex, increasing and $\Psi(0)=0$. In this case Φ is called "Young function" and Ψ the "conjugate function" of Φ. It is not hard to show that q is the exponent of Ψ. Φ is called "co-moderate" if Ψ is moderate and q the "conjugate exponent" of Φ. A pair of Young functions of special importance here is given by

$$\Phi(t) = t\log^+ t,\ \Psi(t) = t1_{[0,1]}(t)+\exp(t-1)\ 1_{]1,\infty[}(t),\ t\in\mathbb{R}_+.$$

For a moderate function Φ and a random variable ξ let

$$\|\xi\|_\Phi = \inf\{\lambda>0:\ E(\Phi(\tfrac{|\xi|}{\lambda}))\le 1\}.$$

$\|\cdot\|_\Phi$ defines a seminorm on $\mathfrak{m}(\mathcal{F},\mathcal{B}(\mathbb{R}))$. By $L^\Phi(\Omega,\mathcal{F},P)$ we denote the linear space of all random variables such that $\|\xi\|_\Phi < \infty$. A process X is said to be "Φ-bounded" or "Φ-integrable", if $\sup_{t\in\mathbb{I}} E(\Phi(|X_t|))<\infty$. In the special case $\Phi(t) = t\log^+ t$ resp. $\Phi(t) = t^2$, $t\in\mathbb{R}_+$, we also call it "L $\log^+$L-bounded" or "L $\log^+$L-integrable" resp. "square integrable".

Besides spaces of random variables, we will encounter topological vector spaces of processes at some places. For a measure m on a σ-algebra $\mathcal{A}\subset[\mathcal{F}\times\mathcal{B}(\mathbb{I})]\vee\mathcal{N}$, we will denote by $L^p(\Omega\times\mathbb{I},\mathcal{A},m)$ the linear space of all $\mathcal{A}$-measurable processes X such that $\int|X|^p\,dm<\infty$, by $L^{p,\infty}(\Omega\times\mathbb{I},\mathcal{A},P)$ correspondingly all processes X such that $\|X\|_{p,\infty} = \|\sup_{t\in\mathbb{I}} |X_t|\,\|_p <\infty$, for $p>0$ resp. $p\ge 0$, topologized by these functionals.

I. Basics; processes depending on a parameter

The set up of the most basic notions and facts about two-parameter processes is the first subject of this chapter. Section 1 contains some crucial results about spaces of regular functions of two variables in which, as will be shown later, $L\log^+L$-bounded martingales have almost all of their sample functions. In section 2, the basic measurability notions for two-parameter processes, progressive measurability, optionality and previsibility, are defined and briefly discussed before the basic processes are introduced and their simplest properties investigated. In the following sections 3 to 6, the theory of two-parameter processes is considered from a purely one-parameter point of view, a two-parameter process as a one-parameter process depending reasonably in a way to be precised on the other parameter. This is the starting point for studying the behavior of two-parameter processes consecutively and independently in the two "time" directions in chapter two. Section 3 presents a stopping notion which is natural for this approach, the "stopping point", and a section theorem for stopping points. The latter is applied in section 4 to prove the main projection and dual projection theorems for processes depending measurably on a real parameter. The important norm inequalities for (dual) projections complete section 4 and enable us to compare the norms of processes with the norms of their projections. The situation in section 5 is somewhat more general: instead of increasing processes we investigate the increasing processes associated with the Doléans function of submartingales, depending reasonably on a parameter. We obtain decomposition theorems and Garsia-Neveu-type norm inequalities for the increasing processes figuring in the decompositions. The question we face in section 6 is: which regularity properties do processes inherit to their respective projections according to section 4? We discuss results for the existence of limits and continuity in the four different quadrants and show by a counterexample that the existence of limits in the left-hand quadrants with respect to the projection direction may not be preserved.

1. Regular functions and their discontinuities

One of the fundamental results of martingale theory states that any L^1-bounded martingale possesses a version whose sample functions are right continuous and possess left limits. The corresponding result for two-parameter martingales will be proved in section 9 and constitutes one of the cornerstones of our analysis. It says that a martingale which is bounded in $L \log^+ L$ has a version which is continuous in the right upper quadrant and possesses limits in the remaining three - a property which will be called "regular" - and stands at the very beginning of our investigation of discontinuities of martingales in chapter 3. To derive this result, it will be helpful to consider "regular" real-valued functions of two variables first and describe their discontinuity sets.

To fix some notation, for $r \in [0,1]$ let

$H_r^+ = \{q \in [0,1]: q > r\}$, $\quad H_r^- = \{q \in [0,1]: q < r\}$,

$\hat{H}_r^+$, $\hat{H}_r^-$ their respective closures in the "right" topology, i.e.

$\hat{H}_r^+ = \{q \in [0,1]: q \geq r\}$, $\quad \hat{H}_r^- = H_r^-$.

For $t \in \Pi$, $j \in \{+,-\}$ set

$$Q_t^{\pm j} = [0,1] \times H_{t_2}^j, \quad Q_t^{j\pm} = H_{t_1}^j \times [0,1],$$

$$Q_t^{\cdot j} = \{t_1\} \times H_{t_2}^j, \quad Q_t^{j\cdot} = H_{t_1}^j \times \{t_2\}.$$

Finally, for $t \in \Pi$, $j \in \{+,-\}^2$ set

$$Q_t^j = H_{t_1}^{j_1} \times H_{t_2}^{j_2}.$$

Correspondingly, $\hat{Q}^{\pm +}$ etc. are defined using $\hat{H}^+$ instead of H^+ etc. We feel free to replace $Q_t^{(+,+)}$ by Q_t^{++}, etc.

The geometrical significance of the precedingly defined sets is clear: $Q_t^{\pm +}$, $Q_t^{\pm -}$ are the intersections with Π of the upper and lower halfplanes in $\mathbb{R}^2$ defined by t, Q_t^{++}, Q_t^{-+}, Q_t^{--}, Q_t^{+-} the intersections with Π of the four quadrants in $\mathbb{R}^2$ defined by t.

If $f: \Pi \to \mathbb{R}$ is a function, $S \in \{Q^{\pm +}, \ldots, \hat{Q}^{--}\}$, set

$$\overline{f}^S(t) = \limsup_{S_t \ni s \to t} f(s), \quad \underline{f}^S(t) = \liminf_{S_t \ni s \to t} f(s), \quad t \in \Pi.$$

Eventually, we use $\overline{f}^{++}$ instead of $\overline{f}^{Q^{++}}$ and f^{++}, if $\overline{f}^{++} = \underline{f}^{++}$, etc.

<u>Definition 1</u>: Let $f: \Pi \to \mathbb{R}$ be a function, $S \in \{Q^{\pm +}, \ldots, \hat{Q}^{--}\}$.

1. f is said to "<u>possess S-limits</u>", if $\overline{f}^S(t) = \underline{f}^S(t)$ for all $t \in \Pi$.
2. f is called "<u>S-continuous</u>", if f possesses S-limits f^S and $f = f^S$.
3. f is called "<u>regular</u>", if f is Q^{++}-continuous and possesses Q^j-limits for $j \in \{+,-\}^2$.

Again, if f possesses Q^{++}-limits, we may write "f^{++}" for "$f^{Q^{++}}$" etc.

The following elementary proposition expresses the fact that regularity might be defined using $\hat{Q}$-limits instead of Q-limits.

<u>Proposition 1</u>: Let $f: \Pi \to \mathbb{R}$ be a function.

1. If f is $Q^{\pm +}$-continuous and possesses $Q^{\pm -}$-limits, then it is $\hat{Q}^{\pm +}$-continuous and possesses $\hat{Q}^{\pm -}$-limits.

   A similar statement holds for $Q^{+\pm}$, $Q^{-\pm}$.
2. If f is Q^{++}-continuous and possesses Q^j-limits for $j \in \{+,-\}^2$, then it is $\hat{Q}^{++}$-continuous and possesses $\hat{Q}^j$-limits for $j \in \{+,-\}^2$.

Proof:

Let us prove 2. for $\hat{Q}^{+-}$-limits. The arguments for $\hat{Q}^{++}$-continuity are easier. Let $t \in \mathbb{I}$ be given. Assume there is a sequence $(t^n)_{n \in \mathbb{N}}$ in $\hat{Q}^{+-}$ which converges to t such that $(f(t^n))_{n \in \mathbb{N}}$ does not converge. Since f possesses Q^{+-}-limits, we can assume that $(t^n)_{n \in \mathbb{N}}$ is on $Q_t^{\cdot -}$. By Q^{++}-continuity, for any $n \in \mathbb{N}$ we may choose $s^n \in Q_t^{+-}$ such that $|s^n - t^n| < \frac{1}{n}$, $|f(s^n) - f(t^n)| < \frac{1}{n}$. Now $(s^n)_{n \in \mathbb{N}}$ is a sequence in Q_t^{+-} converging to t such that $(f(s^n))_{n \in \mathbb{N}}$ does not converge. This conflicts with Q^{+-}-continuity. □

Corollary: Let $f: \mathbb{I} \to \mathbb{R}$ be a function. f is regular iff it is $\hat{Q}^{++}$-continuous and possesses $\hat{Q}^j$-limits for $j \in \{+,-\}^2$. If f is regular, then $f = f^{++}$, $f^{\cdot -} = f^{+-}$, $f^{-\cdot} = f^{-+}$.

The following definition exhibits a simple space of elementary functions which - as will be seen - is dense in the space of regular functions with respect to uniform convergence.

Definition 2: The linear space of functions

$$H = \{ \sum_{J \in \mathbb{J}} a_J 1_J : a_J \in \mathbb{R},\ \mathbb{J} \text{ partition of } \mathbb{I} \text{ by intervals of the form } J = [s,t[\}$$

is called space of "regular elementary functions".

Proposition 2: A function $f: \mathbb{I} \to \mathbb{R}$ is regular iff for any $\varepsilon > 0$ there is $g \in H$ such that $\| f - g \|_\infty < \varepsilon$.

Proof:

1. Let f be regular and $\varepsilon > 0$ be given. For $t \in \mathbb{I}$ choose an interval $J^t = [u^t, v^t[$ such that $t \in (J^t)^\circ$ and $|f(x) - f(y)| < \varepsilon$ for $x, y \in \hat{Q}_t^{++} \cap J^t$. This is possible by proposition 1. Choose $t^1, \ldots, t^m \in \mathbb{I}$ such that $J^{t^1}, \ldots, J^{t^m}$ cover $\mathbb{I}$ and let $\mathbb{J}$ be the partition of $\mathbb{I}$ generated by the intervals $\hat{Q}_{t^i}^{++} \cap J^{t^i}$, $1 \le i \le m$.

   Then the regular elementary function

   $$g = \sum_{J \in \mathbb{J}} f(s^J) 1_J$$

   is suitable.

2. Let f be a function in the closure of $\mathbb{H}$ with respect to $\|\cdot\|_\infty$. Let $t\in\mathbb{I}$, $\varepsilon>0$ be given. Choose $g = \sum_{J\in\mathbb{J}} a_J 1_J \in \mathbb{H}$ such that $\|f-g\|_\infty < \varepsilon/2$. Denote by $\mathbb{K}$ the set of (at least one, at most four) intervals J in $\mathbb{J}$ such that $t\in J^a$. For any $j\in\{+,-\}^2$ there exactly one $J\in\mathbb{K}$ such that $J \cap \hat{Q}^j \neq \emptyset$. In addition, for $j = (+,+)$ we have $t\in J$. Therefore

$$\limsup_{Q_t^j\ni s\to t} f(s) - \liminf_{Q_t^j\ni s\to t} f(s) < \varepsilon \quad \text{for } j\in\{+,-\}^2$$

and in addition

$$\left|\limsup_{Q_t^{++}\ni s\to t} f(s) - f(t)\right| < \varepsilon, \qquad \left|\liminf_{Q_t^{++}\ni s\to t} f(s) - f(t)\right| < \varepsilon.$$

Since ε is arbitrary, the regularity of f follows. □

Proposition 2 provides us with a handy tool to study the discontinuities of regular functions.

<u>Definition 3</u>: Let $f:\mathbb{I}\to\mathbb{R}$ be regular, $t\in\mathbb{I}$.

1. We set $\Delta^1_{t_1} f(\cdot,t_2) = f(t) - f^{-\cdot}(t)$, $\Delta^2_{t_2} f(t_1,\cdot) = f(t) - f^{\cdot -}(t)$,

$$\Delta_t f = f(t) - f^{-\cdot}(t) - f^{\cdot -}(t) + f^{--}(t).$$

2. We call t "<u>0-jump</u>" of f, if $\Delta_t f\neq 0$, "<u>i-jump</u>" of f, if $\Delta^i_{t_i} f(\cdot,t_{\bar{i}})\neq 0$, $i=1,2$.

<u>Remark</u>: Let $J_1 = [0,1/2[$, $J_2 = [1/2,1]$, $J^{++} = J_2\times J_2$, $J^{-+}=J_1\times J_2$, $J^{+-} = J_2\times J_1$, $J^{--} = J_1\times J_1$. Define

$$f = 1_{J^{--}}, \quad g = 2\, 1_{J^{++}} + 1_{J^{+-}} + 1_{J^{-+}}.$$

Then $(1/2,1/2)$ is a 0-jump, but neither a 1-nor a 2-jump of f. It is also a 1- and 2-jump, but no 0-jump of g. These examples show that there are no a priori relations between the different kinds of jumps.

However, proposition 2 tells us that the sets of jumps of regular functions are not too complicated.

<u>Theorem 1</u>: Let $f:\mathbb{I}\to\mathbb{R}$ be a regular function. For $0\le\varepsilon$, $0\le i\le 2$

let $D^i(\varepsilon) = \{t \in \Pi$: t is an i-jump of f of height bigger than $\varepsilon\}$. Then

$D^0(\varepsilon)$ is a finite (countable) set in Π,

$D^1(\varepsilon)$ is contained in a finite (countable) union of vertical line segments of Π,

$D^2(\varepsilon)$ is contained in a finite (countable) union of horizontal line segments of Π,

if $\varepsilon>0$ ($\varepsilon=0$).

Proof:

Since $D^i(0) = \bigcup_{n \in \mathbb{N}} D^i(1/n)$, $0 \leq i \leq 2$, it is enough to consider the case $\varepsilon>0$. In this case, the statement is obviously true for regular elementary functions. By proposition 2, it follows for general regular f. □

If a given regular function f has no 0-jumps, our statements on $D^i(\varepsilon)$, i=1,2, can be made more precise.

Theorem 2: Let $f: \Pi \to \mathbb{R}$ be a regular function. Suppose f has no 0-jumps. Then, with the notations of theorem 1:

$D^1(\varepsilon)$ consists of a countable union of open vertical line segments (and is contained in a finite union of vertical line segments of Π),

$D^2(\varepsilon)$ consists of a countable union of open horizontal line segments (and is contained in a finite union of horizontal line segments of Π),

if $\varepsilon=0$ ($\varepsilon>0$).

Proof:

Since f has no 0-jumps, $\Delta^i_{t_i} f(t_{\bar{i}},\cdot)$ is continuous in $t_{\bar{i}}$, i=1,2. This fact together with theorem 1 implies the desired result. □

Remark: The results of the preceding theorems are well known. See Mazziotto, Szpirglas [26].

We finally single out a special class of regular functions.

Proposition 3: Let $f: \mathbb{I} \to \mathbb{R}$ be a regular function. Then: f is of bounded variation iff there exist increasing regular functions g_1, g_2 such that $f = g_1 - g_2$.

Proof:

Decompose the signed measure associated with f according to Jordan. □

Remarks: 1. For the definition of "bounded variation" for functions of two variables see Clarkson, Adams [15].

2. We could consider "regularity" with respect to the remaining three quadrants, too. But, since martingale theory is not invariant with respect to "time inversion" on any axis, it turns out that there is just one relevant kind of regularity, the one we chose to consider.

2. Measurability and basic processes

In this section of introductory and auxiliary character we define the basic measurability concepts of the theory of two-parameter processes: progressive measurability, optionality and previsibility. For later reference we prove some results concerning the inheritance of measurability properties by processes which are gained by basic operations from measurable processes. We finally introduce increasing processes and martingales.

If one considers two-parameter processes as one-parameter processes depending measurably on the other, the following notions are relevant. Compare Bakry [2] and Meyer [32], [33].

Definition 1: Let i=1,2.

1. Let X be a process on $\Omega\times[0,1]$. X is called "$\mathbb{F}_i$-adapted", if $X_r \in \mathfrak{M}(\mathcal{F}_r^i, \mathcal{B}(\mathbb{R}))$, "$\mathbb{F}_i$-progressively measurable", if $X|_{\Omega\times[0,r[} \in \mathfrak{M}([\mathcal{F}_r^i\times\mathcal{B}([0,r[)]\vee\mathcal{N}, \mathcal{B}(\mathbb{R}))$ for any $r\in[0,1]$. Denote by $\mathfrak{M}_i$ the σ-algebra of $\mathbb{F}_i$-progressively measurable sets.

2. Let X be a process on $\Omega\times\mathbb{I}$. X is called "i-adapted", if $X_t \in \mathfrak{M}(\mathcal{F}_{t_i}^i, \mathcal{B}(\mathbb{R}))$. The σ-algebra $\mathfrak{M}^i = [\mathfrak{M}_i\times\mathcal{B}([0,1])]\vee\mathcal{N}$ is called σ-algebra of "i-progressively measurable sets".

Definition 2: Let i=1,2.

1. The σ-algebra $\mathfrak{G}_i$ of "$\mathbb{F}_i$-optional sets" is generated by the right-continuous, $\mathbb{F}_i$-adapted processes on $\Omega\times[0,1]$ possessing left limits; $\mathfrak{G}^i = [\mathfrak{G}_i\times\mathcal{B}([0,1])]\vee\mathcal{N}$ is called σ-algebra of "i-optional sets".

2. The σ-algebra $\mathcal{P}_i$ of "$\mathbb{F}_i$-previsible sets" is generated by the continuous, $\mathbb{F}_i$-adapted processes on $\Omega\times[0,1]$; $\mathcal{P}^i = [\mathcal{P}_i\times\mathcal{B}([0,1])]\vee\mathcal{N}$ is called σ-algebra of "i-previsible sets".

Before giving "two-parameter" versions of the preceding definitions, we transfer the regularity notions of section 1 to stochastic processes.

Definition 3: Let X be a process on $\Omega\times\mathbb{I}$, $S=Q^{\pm+}, \ldots, \hat{Q}^{--}$. X is said to possess "S-limits", to be "S-continuous", "regular", if for all $\omega\in\Omega$ the function $X(\omega,\cdot)$ has this property.

Definition 4: Let X be a process on $\Omega\times\mathbb{I}$. X is called "adapted", if $X_t \in \mathfrak{M}(\mathcal{F}_t,\mathcal{B}(\mathbb{R}))$, "progressively measurable", if $X|_{\Omega\times[0,t[} \in \mathfrak{M}([\mathcal{F}_t\times\mathcal{B}([0,t[)]\vee\mathcal{N}, \mathcal{B}(\mathbb{R}))$ for any $t\in\mathbb{I}$. Denote by $\mathfrak{M}$ the σ-algebra of progressively measurable sets.

Definition 5:

1. The σ-algebra $\mathfrak{G}$ of "optional sets" is generated by the regular adapted processes on $\Omega\times\mathbb{I}$.

2. The σ-algebra $\mathbb{P}$ of "previsible sets" is generated by the continuous adapted processes on $\Omega\times\mathbb{I}$.

The hierarchy of the different notions of measurability is as follows.

Proposition 1: We have $\mathbb{P}\subset\mathbb{G}\subset\mathbb{M}$, $\mathbb{P}^i\subset\mathbb{G}^i\subset\mathbb{M}^i$ for $i=1,2$.

Proof:

For the "one-directional" results, we refer the reader to Métivier [31], pp. 9-11, where the corresponding one-parameter problem is treated. Note that $\mathbb{P}\subset\mathbb{G}$ is trivial. To prove $\mathbb{G}\subset\mathbb{M}$, let X be adapted and regular. Then X can be approximated by the sequence

$$X_n = \sum_{J\in\mathbb{J}_n} X_{t^J} 1_J, \quad n\in\mathbb{N},$$

if $(\mathbb{J}_n)_{n\in\mathbb{N}}$ is an arbitrary 0-sequence of partitions of $\mathbb{I}$. Now for any $n\in\mathbb{N}$, X_n is clearly progressively measurable, hence so is X. By definition of $\mathbb{G}$, the assertion follows. □

The σ-algebras of previsible sets have particularly simple generators, the "previsible rectangles" in $\Omega\times\mathbb{I}$.

Definition 6: Let $i=1,2$. Then

$\mathbb{R}^i = \{F\times]s,t]: F\in\mathcal{F}^i_{s_i}, s,t\in\mathbb{I}, s\leq t\}$ is called the set of "i-previsible rectangles". Similarly, the set $\mathbb{R}$ of "previsible rectangles" is defined.

Proposition 2: The σ-algebra $\mathbb{P}$ is generated by $\mathbb{R}$, $\mathbb{P}^i$ by $\mathbb{R}^i$, $i=1,2$.

Proof:

We argue for $\mathbb{P}$ and $\mathbb{R}$. Let $R = F\times J\in\mathbb{R}$ be given. Approximate 1_J by an increasing sequence $(f_n)_{n\in\mathbb{N}}$ of continuous functions on $\mathbb{I}$ which are zero on $\overline{[s^J,(1,1)]}$. Then $X_n = 1_F\cdot f_n$, $n\in\mathbb{N}$, defines a sequence of continuous adapted processes which converges to 1_R. Hence $\sigma_{\Omega\times\mathbb{I}}(\mathbb{R})\subset\mathbb{P}$. If, on the other hand, X is a continuous adapted process on $\Omega\times\mathbb{I}$, we may approximate it, given a 0-sequence $(\mathbb{J}_n)_{n\in\mathbb{N}}$ of partitions of $\mathbb{I}$, by

$$X_n = \sum_{J\in \mathbb{J}_n} X_{s^J} 1_J, \quad n\in\mathbb{N}.$$

Obviously, $X_n \in \mathfrak{M}(\sigma_{\Omega\times\mathbb{I}}(\mathcal{R}), \mathcal{B}(\mathbb{R}))$ for all $n\in\mathbb{N}$. By definition of $\mathcal{P}$, the inclusion $\mathcal{P} \subset \sigma_{\Omega\times\mathbb{I}}(\mathcal{R})$ follows and we are done. □

The following important theorem gives some information on the measurability properties of limit processes.

Theorem 1:

1. Let $X \in \mathfrak{M}(\mathfrak{M}^1, \mathcal{B}(\mathbb{R}))$. Then:

   i) $\overline{X}^S, \underline{X}^S \in \mathfrak{M}(\mathfrak{M}^1, \mathcal{B}(\mathbb{R}))$ for all $S = Q^{\pm+}, \ldots, Q^{--}$,

   ii) $\overline{X}^S, \underline{X}^S \in \mathfrak{M}(\mathcal{P}^1, \mathcal{B}(\mathbb{R}))$ for $S = Q^{-\pm}, Q^{-+}, Q^{-\cdot}, Q^{--}$.

A corresponding statement holds with respect to the second parameter.

2. Let $X \in \mathfrak{M}(\mathfrak{M}, \mathcal{B}(\mathbb{R}))$. Then:

   i) $\overline{X}^S, \underline{X}^S \in \mathfrak{M}(\mathfrak{M}, \mathcal{B}(\mathbb{R}))$ for all $S = Q^{\pm+}, \ldots, Q^{--}$,

   ii) $\overline{X}^S, \underline{X}^S \in \mathfrak{M}(\mathcal{P}^1, \mathcal{B}(\mathbb{R}))$ for $S = Q^{-\pm}, Q^{-+}, Q^{-\cdot}$,

   $\overline{X}^S, \underline{X}^S \in \mathfrak{M}(\mathcal{P}^2, \mathcal{B}(\mathbb{R}))$ for $S = Q^{\pm-}, Q^{+-}, Q^{\cdot-}$,

   $\overline{X}^{--}, \underline{X}^{--} \in \mathfrak{M}(\mathcal{P}, \mathcal{B}(\mathbb{R}))$.

   If in addition $X \in \mathfrak{M}(\mathcal{P}^1, \mathcal{B}(\mathbb{R}))$,

   iii) $\overline{X}^{\cdot-}, \underline{X}^{\cdot-} \in \mathfrak{M}(\mathcal{P}^1\cap\mathcal{P}^2, \mathcal{B}(\mathbb{R}))$.

A corresponding statement holds with respect to the second parameter.

Proof:

We argue for the more difficult second part of the assertion and concentrate on upper limits.

1. If for $t\in\mathbb{I}$ the process $X|_{\Omega\times[0,t[}$ is $[\mathcal{F}_t\times\mathcal{B}([0,t[)]\vee\mathcal{N}$-measurable, the same is obviously true for $\overline{X}^S|_{\Omega\times[0,t[}$, $S = Q^{\pm+}, \ldots, Q^{--}$. This implies i).

2. The arguments for the other cases being similar, we prove ii) for $\overline{X}^{--}$. Fix a 0-sequence $(\mathbb{J}_n)_{n\in\mathbb{N}}$ of partitions of $\mathbb{I}$ and for $n,m\in\mathbb{N}$, $n\le m$, set

$$X^{n,m} = \sum_{J\in\mathbb{J}_n} \sum_{\mathbb{J}_m \ni H\subset J} \sup_{s\in J\cap[0,s^H[} X_s \cdot 1_H.$$

Then, on one hand, $X^{n,m}$ is $\mathfrak{P}$-measurable for all $n,m \in \mathbb{N}$, $n \leq m$ (proposition 2), and on the other hand, $\inf_{n \in \mathbb{N}} \sup_{m \geq n} X^{n,m} = \overline{X}^{--}$.

3. If $X \in \mathfrak{M}(\mathfrak{P}^1, \mathfrak{B}(\mathbb{R}))$, by an argument like in 2., $\overline{X}^{\cdot -} \in \mathfrak{M}(\mathfrak{P}^1, \mathfrak{B}(\mathbb{R}))$. But by ii$_)$, $\overline{X}^{\cdot -} \in \mathfrak{M}(\mathfrak{P}^2, \mathfrak{B}(\mathbb{R}))$. This implies iii$_)$. □

Remark: As will be seen in section 10, $\mathfrak{P}^1 \cap \mathfrak{P}^2 = \mathfrak{P}$, by conditional independence of $\mathbb{F}_1$ and $\mathbb{F}_2$. This gives iii$_)$ of theorem 1 its final form.

We next note some measurability properties which the interiors and closures of measurable sets inherit.

Theorem 2: Let $S \subset \Omega \times \amalg$. Then:

$S \in \mathfrak{M}$ implies S^a, $S^o \in \mathfrak{M}$. An analogous statement holds with $\mathfrak{M}^i$, $i=1,2$.

Proof:

For $S \in \mathfrak{M}$ we show $S^o \in \mathfrak{M}$. The corresponding result for the closure is obtained by considering complements. Let $\mathfrak{B}$ be a countable base of $\amalg$. For $B \in \mathfrak{B}$ let

$$S_B = \{\omega \in \Omega : B \subset S_\omega\}.$$

Then

$$S^o = \bigcup_{B \in \mathfrak{B}} (S_B \times B).$$

Moreover, if $B \subset [0,t[$ for some $t \in \amalg$, we have

$$\overline{S_B} = \{\omega \in \Omega : B \cap \overline{S_\omega} \neq \emptyset\} = \pi_\Omega(\Omega \times B \cap \overline{S}) \in \mathcal{F}_t,$$

since $\mathcal{F}_t$ is complete (see Dellacherie, Meyer [17], p.68, p.92). This implies $S^o \in \mathfrak{M}$. □

We will not have full information about the relationships between "measurability" and "i-measurability", i=1,2, before section 10. So far, we can only state the following.

Remark: For any $t \in \amalg$, we have $\mathcal{F}_t = \mathcal{F}^1_{t_1} \cap \mathcal{F}^2_{t_2}$. This easily implies that a process is adapted iff it is 1- and 2-adapted and that $\mathfrak{M} = \mathfrak{M}^1 \cap \mathfrak{M}^2$. Similar equations hold for the optional and previsible sets and will be derived as by-products in section 10.

From the definitions, we have the trivial inclusions $\mathcal{P} \subset \mathcal{P}^1 \cap \mathcal{P}^2$ and $\mathcal{O} \subset \mathcal{O}^1 \cap \mathcal{O}^2$.

Let us now consider the most important types of processes.

Definition 7: A process A on $\Omega \times \Pi$ is called "increasing", if

i) $\Delta_J A \geq 0$ for any $J \in \mathcal{J}$,

ii) A is regular.

Definition 8: Let i=1,2.

1. An i-adapted process M is called "i-(sub-)martingale", if M_t is integrable for all $t \in \Pi$ and $E(\Delta_J M \mid \mathcal{F}^i_{s_i J}) \overset{(\geq)}{=} 0$ for any $J \in \mathcal{J}$.

2. An adapted process M is called "weak (sub-)martingale", if M_t is integrable for all $t \in \Pi$ and $E(\Delta_J M \mid \mathcal{F}_{s J}) \overset{(\geq)}{=} 0$ for any $J \in \mathcal{J}$.

3. An adapted process M is called "(sub-)martingale", if M_t is integrable for all $t \in \Pi$ and $E(M_t \mid \mathcal{F}_s) \overset{(\geq)}{=} M_s$ for $s \leq t$.

Remarks: Let i=1,2.

1. A process M is an i-martingale iff $M_{(t_i,\cdot)}$ is an $\mathbb{F}_{\bar{i}}$ - martingale for all $t_i \in [0,1]$. An adapted i-martingale is a weak martingale.

2. Any martingale is a weak martingale. A process M is a martingale iff it is both a 1- and a 2-martingale. This fact is due to the property of conditional independence of $\mathbb{F}_1$ and $\mathbb{F}_2$.

There are important relationships between processes and measures on σ-algebras in $\mathcal{F} \times \mathcal{B}(\Pi)$, which, like in the theory of one-parameter processes, will be particularly useful when we investigate projection and decomposition theorems in later sections.

Definition 9: A finite measure m on a σ-algebra $\mathcal{N} \subset \mathcal{H} \subset (\mathcal{F} \times \mathcal{B}(\Pi)) \vee \mathcal{N}$ is called "admissible", if m vanishes on $\mathcal{N}$.

Proposition 3: The correspondence between admissible measures m on $[\mathcal{F} \times \mathcal{B}(\Pi)] \vee \mathcal{N}$ and integrable increasing processes A on $\Omega \times \Pi$ such that

$$m(S) = E(\int_{\Pi} 1_S \, dA) \qquad \text{for } S \in [\mathcal{F} \times \mathcal{B}(\Pi)]\vee\mathcal{N} \ ,$$

is one-to-one.

Proof:

Let m be an admissible measure on $[\mathcal{F} \times \mathcal{B}(\Pi)]\vee\mathcal{N}$. For $t \in \Pi$, define

$$m_t(B) = m(B \times [0,t]), \qquad B \in \mathcal{F}.$$

Then m_t is a measure on $\mathcal{F}$ which is clearly P-continuous since m is admissible. Let V_t be a Radon-Nikodym-derivative of m_t with respect to P. The process V can be chosen increasing on $\mathbb{Q}^2 \cap \Pi$. Define the process A by

$$A_t := \lim_{\substack{\mathbb{Q}^2 \ni s \to t \\ s \geq t}} V_s, \qquad t \in \Pi.$$

A is an integrable increasing process which clearly satisfies the desired relation. The rest of the assertion is standard. □

A bit more generally, one can associate with each weak or i-submartingale a non-negative set function on the algebras generated by $\mathcal{R}$ or $\mathcal{R}^i$, i=1,2.

Definition 10: Let i=1,2, M an i-submartingale (a weak submartingale). The extension m_M^o of the weakly additive set function

$$m_M^o(F \times J) = E(1_F \cdot \Delta_J M), \qquad F \times J \in \mathcal{R}^i \ (\mathcal{R})$$

to the algebra generated by $\mathcal{R}^i$ ($\mathcal{R}$), is called "Doléans set function" of M. A possibly existing finite σ-additive extension of m_M^o to $\mathcal{P}^i$ ($\mathcal{P}$) is called "Doléans measure" of M and denoted by m_M.

Remarks: 1. Let i=1,2. An i-submartingale (weak submartingale) M is an i-martingale (weak martingale) iff $m_M^o = 0$. Martingales cannot be characterized in this way (see Merzbach [27], pp. 37,38).

2. If A is an integrable increasing process, the measure existing according to proposition 3 is even a unique extension of the Doléans measure to $[\mathcal{F} \times \mathcal{B}(\Pi)]\vee\mathcal{N}$. We denote it also by m_A.

3. Extension theorems for Doléans set functions will be the

essence of decomposition theorems for submartingales.

Another type of important measures associated with processes is the stochastic measure or stochastic integral which, in the case of an increasing process, is just a pointwise Stieltjes-integral.

Definition 11: Let A be an increasing process, Y a process on $\Omega \times \mathbb{I}$ such that for all $\omega \in \Omega$ the integral $\int_{\mathbb{I}} Y.(\omega)\, dA.(\omega)$ exists. Then

$$Y \cdot A = \int_{\mathbb{I}} Y\, dA \quad \text{and} \quad Y \cdot A. = \int_{[0,\cdot]} Y\, dA$$

are called "integral of Y w.r.t. A" respectively "integral process of Y w.r.t. A". For $S \subset \Omega \times \mathbb{I}$, the process $1_S \cdot A.$ is also denoted by $A(S)$.

Integral processes inherit some measurability properties from integrator and integrated process.

Proposition 4: Let A be an increasing process, Y a process on $\Omega \times \mathbb{I}$ such that $Y \cdot A.$ exists, $i=1,2$. Then:

1. $A \in \mathfrak{M}(\mathfrak{G}^i, \mathfrak{B}(\mathbb{R}))$, $Y \in \mathfrak{M}(\mathfrak{M}^i, \mathfrak{B}(\mathbb{R}))$ implies $Y \cdot A. \in \mathfrak{M}(\mathfrak{G}^i, \mathfrak{B}(\mathbb{R}))$,
2. $A \in \mathfrak{M}(\mathfrak{P}^i, \mathfrak{B}(\mathbb{R}))$, $Y \in \mathfrak{M}(\mathfrak{P}^i, \mathfrak{B}(\mathbb{R}))$ implies $Y \cdot A. \in \mathfrak{M}(\mathfrak{P}^i, \mathfrak{B}(\mathbb{R}))$.

Similar statements hold with $\mathfrak{G}^i$, $\mathfrak{M}^i$, $\mathfrak{P}^i$ replaced by $\mathfrak{G}$, $\mathfrak{M}$, $\mathfrak{P}$.

Proof:

If $Y \in \mathfrak{M}(\mathfrak{M}, \mathfrak{B}(\mathbb{R}))$, it is obvious that $Y \cdot A. \in \mathfrak{M}(\mathfrak{M}, \mathfrak{B}(\mathbb{R}))$. Moreover, $Y \cdot A.$ is regular. Hence it is optional.

If $F \times J$, $G \times K \in \mathfrak{R}$, we have $1_{F \cap G}\, 1_{J \cap K \cap [0,\cdot]} \in \mathfrak{M}(\mathfrak{P}, \mathfrak{B}(\mathbb{R}))$.

A monotone class argument, supported by proposition 2, gives $1_F\, \Delta_{J \cap [0,\cdot]} A \in \mathfrak{M}(\mathfrak{P}, \mathfrak{B}(\mathbb{R}))$, if A is previsible. Another monotone class argument yields the desired measurability result. □

To study stochastic integrals with respect to martingales M, as usual, requires some deeper knowledge. For example, we need a decomposition theorem for the submartingale M^2. We will come back to this problem in section 11. The principal aims of this book are structure theorems for square integrable martingales and their quadratic variations. For these purposes we need a notion of "orthogonality" of square integrable martingales and of quadratic variations.

Definition 12: Square integrable martingales M and N are said to be "orthogonal", if MN is a weak martingale.

Definition 13: Two processes X and Y are said to have "orthogonal variation", if for any 0-sequence $(\mathbb{I}_n)_{n \in \mathbb{N}}$ of partitions of $\mathbb{I}$

$$\sum_{J \in \mathbb{I}_n} |\Delta_J X \, \Delta_J Y| \underset{n \to \infty}{\to} 0 \quad \text{in} \quad L^0(\Omega, \mathcal{F}, P).$$

We will show in section 8 that square integrable martingales of orthogonal variation are orthogonal.

3. One-directional stopping; a section theorem

A concept which causes considerable problems in the theory of multi-parameter processes is the appropriate random choice of objects in parameter space (like points) given the information "up to them". For purely geometric reasons points are inadequate to define something like "past" or "future". However, if one looks in one direction only and all the information depending on the other parameter is always available, they prove to be useful. To visualize a "stopping point", consider the family of all lines vertical to the (axial parallel) direction in which you look, parametrized by its time variable. Make a random choice of one of these lines using only information which is available on the lines "before". Finally pick a point on the chosen line, again on this information basis. In this study stopping points are used for projection theorems and regularity theorems mainly.

We start by proving that, viewed in i-direction, i-progressively measurable sets possess stopping points of first entrance. We then establish the most important section theorem presented here, a one-directional optional and previsible section theorem for stopping points. Still following Bakry [2], [3], we conclude this short section by showing that stopping points possess "fundamental systems" of previsible neighborhoods in certain quadrants and half planes.

Proposition 1: Let $S \in \mathfrak{M}^1$. Then there exists an $\mathbb{F}_1$-stopping time T_1, and a random variable $T_2 \in \mathfrak{M}(\mathcal{F}^1_{T_1}, \mathcal{B}(\mathbb{R}))$ such that

i) $T_1 = \inf\{t_1 \in [0,1]: \Omega \times \{t_1\} \times [0,1] \cap S \neq \emptyset\}$,

ii) $[(T_1,T_2)] \subset S^a$ on $\{[T_1] \times [0,1] \cap S^a \neq \emptyset\}$,

If $S \in \mathcal{P}^1$ and $S=S^a$, T_1 can be chosen previsible and $T_2 \in \mathfrak{M}(\mathcal{F}^1_{T_1-}, \mathcal{B}(\mathbb{R}))$. An analogous statement holds for the second parameter.

Proof:

By theorem (2.2) we may assume $S = S^a$. Define T_1 by i). Then for $r \in [0,1]$

$$\{T_1 < r\} = \pi_\Omega(\Omega \times [0,r[\times [0,1] \cap S) \in \mathcal{F}^1_r$$

by the completeness of $\mathbb{F}_1$ (see Dellacherie, Meyer [17], p.68). Let

$$C = \{(\omega,t_2) \in \Omega \times [0,1]: (\omega,T_1(\omega),t_2) \in S\}.$$

Then $C \in \mathcal{F}^1_{T_1} \times \mathcal{B}([0,1])$ and by the classical section theorem (Dellacherie, Meyer [17], p.103) there is an $\mathcal{F}^1_{T_1}$-measurable random variable T_2 such that ii) holds.

If $S \in \mathcal{P}^1$, T_1 is previsible (see Métivier [31], p.17) and $C \in \mathcal{F}^1_{T_1-} \times \mathcal{B}([0,1])$.

Therefore, T_2 can be chosen in $\mathfrak{M}(\mathcal{F}^1_{T_1-}, \mathcal{B}(\mathbb{R}))$. □

The object called "stopping point" in the following definition appears, at first sight, more geometrical than the one that showed up in proposition 1. But we will prove shortly that they are the same.

Definition 1: Let $i=1,2$. A set $T \subset \Omega \times \Pi$ is called "i-stopping point", if for $\omega \in \Omega$ the set T_ω contains exactly one element and $T \in \mathcal{G}^i$. By T_1, T_2 we denote the mappings from Ω to $[0,1]$ such that $[(T_1,T_2)]=T$. An i-stopping point $T \in \mathcal{P}^i$ is simply called "previsible i-stopping point".

Proposition 2: Let $i=1,2$, $T \subset \Omega \times \Pi$ a set such that $|T_\omega| = 1$ for $\omega \in \Omega$. Then T is a (previsible) i-stopping point iff T_i is a (previsible) $\mathbb{F}_i$-stopping time and $T_{\bar{i}} \in \mathfrak{m}(\mathcal{F}^i_{T_i}, \mathcal{B}(\mathbb{R}))$ $(\mathfrak{m}(\mathcal{F}^i_{T_i-}, \mathcal{B}(\mathbb{R})))$.

Proof:

We argue for the previsible case if $i=1$. Let T_1 be a previsible $\mathbb{F}_1$-stopping time, $T_2 \in \mathfrak{m}(\mathcal{F}^1_{T_1-}, \mathcal{B}(\mathbb{R}))$. For $m \in \mathbb{N}$, $1 \le i \le m$, let

$$F_{i,m} = \{\tfrac{i-1}{m} < T_2 \le \tfrac{i}{m}\}, \qquad T_2^m = \sum_{1 \le i \le m} \tfrac{i-1}{m} 1_{F_{i,m}}.$$

Then for $n,m \in \mathbb{N}$

$$S_{n,m} = \{(\omega,t) \in \Omega \times \Pi : T_1(\omega) \le t_1 < T_1(\omega) + \tfrac{1}{n},\ T_2^m(\omega) \le t_2 < T_2^m(\omega) + \tfrac{1}{n}\}$$

$$= \bigcup_{1 \le i \le m} [(T_1)_{F_{i,m}}, (T_1)_{F_{i,m}} + \tfrac{1}{n}[\times [\tfrac{i-1}{m}, \tfrac{i-1}{m} + \tfrac{1}{n}[\in \mathcal{P}_1 \times \mathcal{B}([0,1]) \subset \mathcal{P}^1$$

(see Dellacherie, Meyer [17],p.205). Now obviously $[T] = \lim_{n\to\infty} \lim_{m\to\infty} S_{n,m}$. This implies $[T] \in \mathcal{P}^1$. The opposite implication follows from proposition (2.1) and proposition 1. □

If a given i-previsible set is not necessarily closed, we can still find a previsible i-stopping point which is a "section" of it - up to a remainder of arbitrarily small probability.

Theorem 1: Let $i=1,2$, $S \in \mathcal{G}^i(\mathcal{P}^i)$, $\varepsilon > 0$.

Then there exists a (previsible) i-stopping point T such that

i) $T = (1,1)$ on $\{T \not\subset S\}$,

ii) $P(\pi_\Omega(S) \cap \pi_\Omega(\bar{S} \cap T)) < \varepsilon$.

Proof:

We argue for the 1-optional case. The proof in the previsible case is almost identical. The set $\pi_{\Omega,1}(S)$ is $\mathcal{G}_1$-analytic, according to Dellacherie, Meyer [17], p.68. Since the classical optional section theorem (Dellacherie, Meyer [17], p.219) extends, with the same proof, to $\mathcal{G}_1$-analytic sets, we can choose an $\mathbb{F}_1$-stopping time T_1 such that

$T_1 = 1$ on $\{[T_1] \not\subset \pi_{\Omega,1}(S)\}$ and

$P(\pi_\Omega(S) \cap \{[T_1] \not\subset \pi_{\Omega,1}(S)\}) < \varepsilon$.

Now let

$C = \{(\omega,t_2) \in \Omega \times [0,1]: (\omega,T_1(\omega),t_2) \in S\}$

and choose T_2 like in the proof of proposition 1. Note that it is possible to have $T_2 = 1$ on $\{[T_2] \not\subset C\}$. This completes the proof. □

The following propositions show how to define arbitrarily small previsible neighborhoods for a given stopping point in the quadrants and half planes it defines. The left hand quadrants with respect to the projection direction are exceptional.

Proposition 3: Let T be a 1-stopping point.

1. There is a sequence $(R_n)_{n\in\mathbb{N}}$ of 1-previsible sets such that

i) $[(R_n)_\omega]^a$ is a neighborhood of T_ω in $[Q^{+\pm}_{T_\omega}]^a$ for $\omega\in\Omega$, $n\in\mathbb{N}$,

ii) $\operatorname{diam}(R_n) \underset{n\to\infty}{\to} 0$.

2. If T is previsible, there is a sequence $(S_n)_{n\in\mathbb{N}}$ of 1-previsible sets such that

i) $[(S_n)_\omega]^a$ is a neighborhood of T_ω in $[Q^{-\pm}_{T_\omega}]^a$ for $\omega\in\Omega$, $n\in\mathbb{N}$,

ii) $\operatorname{diam}(S_n) \underset{n\to\infty}{\to} 0$.

An analogous statement holds for 2-stopping points, with $Q^{\pm+}$, $Q^{\pm-}$ instead of $Q^{+\pm}$, $Q^{-\pm}$.

Proof:

1. For $n \in \mathbb{N}$ put

$$R_n = \{(\omega,t) \in \Omega \times \Pi : T_1(\omega) < t_1 \leq T_1(\omega)+\frac{1}{n},\ T_2(\omega)-\frac{1}{n} \leq t_2 < T_2(\omega)+\frac{1}{n}\}.$$

Obviously, $(R_n)_{n \in \mathbb{N}}$ fulfills $i_)$ and $ii_)$. To see that R_n is 1-previsible, proceed like in the proof of proposition 2.

2. Suppose T to be 1-previsible. Choose a sequence $(T_1^n)_{n \in \mathbb{N}}$ of previsible $\mathbb{F}_1$-stopping times which announces T_1 (see Dellacherie, Meyer [17], p.211). For $n,m \in \mathbb{N}$, $n \leq m$, let

$$\overline{T}^{n,m} = \sup_{n \leq k \leq m} E(T_2 | \mathcal{F}^1_{T_1^k}), \qquad \underline{T}^{n,m} = \inf_{n \leq k \leq m} E(T_2 | \mathcal{F}^1_{T_1^k}),$$

$$\overline{T}^{n} = \sup_{n \leq k} E(T_2 | \mathcal{F}^1_{T_1^k}), \qquad \underline{T}^{n} = \inf_{n \leq k} E(T_2 | \mathcal{F}^1_{T_1^k}),$$

$$S_{n,m} = \{(\omega,t) \in \Omega \times \Pi : T_1^m(\omega) < t_1 \leq T_1^{m+1}(\omega),\ \underline{T}^{n,m}(\omega)-\frac{1}{n} \leq t_2 < \overline{T}^{n,m}(\omega)+\frac{1}{n}\}.$$

Again like in the proof of proposition 2, $S_{n,m}$ turns out to be 1-previsible for $n,m \in \mathbb{N}$, $n \leq m$. Now let for $n \in \mathbb{N}$

$$S_n = \bigcup_{n \leq m} S_{n,m}.$$

Since $T_2 \in \mathfrak{m}(\mathcal{F}^1_{T_1-}, \mathcal{B}(\mathbb{R}))$ and $\bigvee_{n \in \mathbb{N}} \mathcal{F}^1_{T_1^n} = \mathcal{F}^1_{T_1-}$, we have $\overline{T}^n \downarrow T_2$, $\underline{T}^n \uparrow T_2$.

Therefore, $i_)$ and $ii_)$ are satisfied. □

Proposition 4: Let T be a 1-stopping point.

1. There is a sequence $(R_n)_{n \in \mathbb{N}}$ of 1-previsible sets such that

   $i_)$ $[(R_n)_\omega]^a$ is a neighborhood of T_ω in $[Q^{++}_{T_\omega}]^a$ for $\omega \in \Omega$, $n \in \mathbb{N}$,

   $ii_)$ $\operatorname{diam}(R_n) \underset{n \to \infty}{\to} 0$.

2. There is a sequence $(S_n)_{n \in \mathbb{N}}$ of 1-previsible sets such that

   $i_)$ $[(S_n)_\omega]^a$ is a neighborhood of T_ω in $[Q^{+-}_{T_\omega}]^a$ for $\omega \in \Omega$, $n \in \mathbb{N}$,

   $ii_)$ $\operatorname{diam}(S_n) \underset{n \to \infty}{\to} 0$.

An analogous statement holds for 2-stopping points, with Q^{+-} replaced by Q^{-+}.

Proof:

We only have to "half" the sequence of sets in proposition 3,1.

For $n \in \mathbb{N}$, take

$$R_n = \{(\omega,t) \in \Omega \times \mathrm{II}: T_1(\omega) < t_1 \leq T_1(\omega)+\frac{1}{n},\ T_2(\omega) \leq t_2 < T_2(\omega)+\frac{1}{n}\},$$

$$S_n = \{(\omega,t) \in \Omega \times \mathrm{II}: T_1(\omega) < t_1 \leq T_1(\omega)+\frac{1}{n},\ T_2(\omega)-\frac{1}{n} < t_2 \leq T_2(\omega)\}. \ \square$$

Remark: Given a previsible 1-stopping point, a statement corresponding to proposition 3,2. with Q^{-+} and Q^{--} is not necessarily true. This, as will be seen later, complicates the proof of the existence of regular versions for martingales considerably.

4. One-directional projection theorems

To recall the problem we face now in a one-parameter setting, let X be a bounded measurable process, A an integrable increasing process, both indexed by [0,1], like a filtration with respect to which optional and previsible sets are defined. The classical projection theorems claim that there is a unique optional (previsible) process Y, which, viewed from the optional (previsible) sets, looks the same as X, i.e. for any optional (previsible) increasing process C, the integrals of X and Y with respect to m_C coincide. Dually, there is a unique optional (previsible) increasing process B such that m_A and m_B cannot be distinguished by optional (previsible) sets.

In this section we will prove (dual) projection theorems for projections in one time direction. The problem of projecting is therefore the same as in the classical theory: we will roughly apply the above mentioned theorems for every value the second parameter can take. The difficulty we have to overcome is to keep the dependence on this one measurable. Once this is done, also the norm inequalities which compare (Orlicz) norms of processes with norms of their projections will be derived by essentially classical arguments. Projection theorems are at the basis of all decomposition theorems and therefore vital to our main results. For most of the material of this section we follow Bakry [2], Doléans, Meyer [19], and Meyer [32].

Theorem 1: Let $i=1,2$, X a bounded process in $\mathfrak{M}([\mathcal{F}\times\mathcal{B}(\Pi)]\vee\mathcal{N},\mathcal{B}(\mathbb{R}))$. Then there exists a unique process ${}^{\gamma_i}X$ $({}^{\pi_i}X) \in \mathfrak{M}(\mathcal{G}^i, \mathcal{B}(\mathbb{R}))$ $(\mathfrak{M}(\mathcal{P}^i, \mathcal{B}(\mathbb{R})))$ such that for any integrable increasing process $A \in \mathfrak{M}(\mathcal{G}^i, \mathcal{B}(\mathbb{R}))$ $(\mathfrak{M}(\mathcal{P}^i, \mathcal{B}(\mathbb{R})))$

$$\int_{\Omega\times\Pi} X \, dm_A = \int_{\Omega\times\Pi} {}^{\gamma_i}X \, dm_A \qquad \left(\int_{\Omega\times\Pi} {}^{\pi_i}X \, dm_A\right).$$

Proof:

The arguments for the optional and previsible cases being parallel, we concentrate on the first and assume $i=1$.

1. Let $X = 1_{F\times K}$, where $F\in\mathcal{F}$, $K = [s,t[$ an interval in Π. Let M be a right-continuous version of the martingale $E(1_F|\mathcal{F}^1_{\cdot})$ possessing left limits (see Dellacherie [16], p. 98). Then $Y=M1_K\in\mathfrak{M}(\mathcal{G}^1,\mathcal{B}(\mathbb{R}))$, since it is 1-adapted and regular, and for any 0-sequence $(\mathbb{J}_n)_{n\in\mathbb{N}}$ of partitions of Π, any integrable increasing $A\in\mathfrak{M}(\mathcal{G}^1, \mathcal{B}(\mathbb{R}))$ we have

$$\begin{aligned}\int_{\Omega\times\Pi} X \, dm_A &= \lim_{n\to\infty} E\Big(\sum_{J\in\mathbb{J}_n} 1_F \, \Delta_{J\cap K}A\Big)\\ &= \lim_{n\to\infty} E\Big(\sum_{J\in\mathbb{J}_n} M_{t_1^J} \, \Delta_{J\cap K}A\Big)\\ &= \int_{\Omega\times\Pi} Y \, dm_A.\end{aligned}$$

This establishes "existence" for indicators of measurable rectangles. A linearity and monotone class argument completes the existence proof.

2. Assume $Y, Z \in \mathfrak{M}(\mathcal{G}^1, \mathcal{B}(\mathbb{R}))$ fulfill the asserted equation. Let $S = \{Y-Z>0\}$. Using theorem (3.1) choose, for any given $\varepsilon>0$, a 1-stopping point T such that

(4.1) $T = (1,1)$ on $\{T \not\leq S\}$ and

(4.2) $P(\pi_\Omega(S) \cap \{T \not\leq S\}) < \varepsilon$.

Consider the integrable increasing process $A = 1_{\pi_\Omega(S)} 1_{[T,(1,1)]}$.
By (4.1), $A \in \mathfrak{M}(\mathcal{G}^1, \mathcal{B}(\mathbb{R}))$. Therefore

$$E(1_{\pi_\Omega(S)} (Y_T - Z_T)) = \int_{\Omega \times \Pi} (Y-Z)\, dm_A$$
$$= 0.$$

Hence by (4.2) $P(\pi_\Omega(S)) < \varepsilon$. But $\varepsilon > 0$ was arbitrary. Switch the roles of Y and Z in the arguments just given to obtain uniqueness. □

Definition 1: Let $i=1,2$, $X \in \mathfrak{M}([\mathcal{F} \times \mathcal{B}(\Pi)] \vee \mathcal{N}, \mathcal{B}(\mathbb{R}))$ a bounded process. The process ${}^{\gamma_i}X$ (${}^{\pi_i}X$) according to theorem 1 is called "i-optional (i-previsible) projection" of X.

Remarks: 1. Optional and previsible projections can be defined for bigger classes of measurable processes, for example for all nonnegative measurable processes.

2. According to part 1 of the proof of theorem 1, for bounded $X \in \mathfrak{M}(\mathcal{F}, \mathcal{B}(\mathbb{R}))$ we have ${}^{\gamma_1}X = M$, ${}^{\pi_1}X = M^{-\cdot}$, where M is a right-continuous version of the martingale $E(X|\mathcal{F}^1_{\cdot})$ possessing left limits (see also Dellacherie [16], p. 96). A similar statement holds for the second parameter.

Before we are able to prove the analogue of theorem 1 for dual projections, we need to characterize those integrable increasing processes which cannot "distinguish" between measurable processes and their respective projections.

Proposition 1: Let $i=1,2$, A an integrable increasing process. Then $A \in \mathfrak{M}(\mathcal{G}^i, \mathcal{B}(\mathbb{R}))$ ($\mathfrak{M}(\mathcal{P}^i, \mathcal{B}(\mathbb{R}))$) iff for any bounded $X \in \mathfrak{M}([\mathcal{F} \times \mathcal{B}(\Pi)] \vee \mathcal{N}, \mathcal{B}(\mathbb{R}))$

$$\int_{\Omega \times \Pi} X\, dm_A = \int_{\Omega \times \Pi} {}^{\gamma_i}X\, dm_A \quad \left(\int_{\Omega \times \Pi} {}^{\pi_i}X\, dm_A \right).$$

Proof:

Again, we concentrate on the optional case and assume $i=1$.

If $A \in \mathfrak{M}(\mathfrak{G}^1, \mathcal{B}(\mathbb{R}))$, the asserted equation is fulfilled by theorem 1. To prove the converse, assume the equation is fulfilled and fix $r \in [0,1]$. For bounded measurable processes X on $\Omega \times [0,1]$ denote by ${}^{\gamma}X$ the usual optional projection with respect to $\mathbb{F}_1$. Let such a process X be given. Then, by hypothesis

$$E(\int_{[0,1]} X \, dA_{(.,r)}) = \int_{\Omega \times \Pi} X \, 1_{[0,r]} \, dm_A = \int_{\Omega \times \Pi} {}^{\gamma_1}[X \, 1_{[0,r]}] \, dm_A$$

$$= \int_{\Omega \times \Pi} {}^{\gamma}X \, 1_{[0,r]} \, dm_A = E(\int_{[0,1]} {}^{\gamma}X \, dA_{(.,r)}).$$

Therefore, the corresponding one-parameter result (see Dellacherie, Meyer [18], p. 136) applies and yields that $A_{(.,r)}$ is $\mathbb{F}_1$-optional. This clearly implies $A \in \mathfrak{M}(\mathfrak{G}^1, \mathcal{B}(\mathbb{R}))$. □

<u>Theorem 2</u>: Let i=1,2, A an integrable increasing process. Then there exists a unique integrable increasing process A^{γ_i} $(A^{\pi_i}) \in \mathfrak{M}(\mathfrak{G}^i, \mathcal{B}(\mathbb{R}))$ $(\mathfrak{M}(\mathcal{P}^i, \mathcal{B}(\mathbb{R})))$ such that for any bounded $X \in \mathfrak{M}(\mathfrak{G}^i, \mathcal{B}(\mathbb{R}))$ $(\mathfrak{M}(\mathcal{P}^i, \mathcal{B}(\mathbb{R})))$

$$\int_{\Omega \times \Pi} X \, dm_A = \int_{\Omega \times \Pi} X \, dm_{A^{\gamma_i}} \quad (\int_{\Omega \times \Pi} X \, dm_{A^{\pi_i}}).$$

<u>Proof</u>:

To prove the asserted existence result for A^{γ_i}, we have to construct an admissible measure n such that for any bounded $X \in \mathfrak{M}([\mathcal{F} \times \mathcal{B}(\Pi)] \vee \mathcal{N}, \mathcal{B}(\mathbb{R}))$

$$\int_{\Omega \times \Pi} {}^{\gamma_i}X \, dm_A = \int_{\Omega \times \Pi} {}^{\gamma_i}X \, dn = \int_{\Omega \times \Pi} X \, dn.$$

This follows from proposition 1 and proposition (2.3), if we define

$$n(S) = \int_{\Omega \times \Pi} {}^{\gamma_i}(1_S) \, dm_A, \quad S \in [\mathcal{F} \times \mathcal{B}(\Pi)] \vee \mathcal{N}.$$

A monotone class argument reveals that n is appropriate and existence follows. But any measure belonging to an integrable increasing process in $\mathfrak{M}(\mathfrak{G}^i, \mathcal{B}(\mathbb{R}))$ which fulfills the asserted equation must also satisfy the defining equation of n. This entails uniqueness. □

<u>Definition 2</u>: Let i=1,2, A an integrable increasing process.

The integrable increasing process A^{γ_i} (A^{π_i}) according to theorem 2 is called "<u>dual i-optional (i-previsible) projection</u>" of A.

Norm inequalities for dual projections follow immediately from their one-parameter counterparts.

<u>Theorem 3</u>: Let i=1,2, A an integrable increasing process, Φ a moderate function. Then

$$\| A^{\alpha_i}_{(1,1)} \|_\Phi \leq p \|A_{(1,1)}\|_\Phi \quad , \alpha=\gamma,\pi.$$

In particular, for $q \geq 1$

$$\| A^{\alpha_i}_{(1,1)} \|_q \leq q \| A_{(1,1)} \|_q \quad , \alpha=\gamma,\pi.$$

<u>Proof</u>:

Since A is increasing in both directions, the corresponding one-parameter inequalities can be applied (see Dellacherie, Meyer [18], p. 183). □

Norm inequalities for projections extend Doob's inequality and require some more work.

<u>Theorem 4</u>: Let i=1,2, $X \in \mathfrak{m}([\mathcal{F}\times\mathcal{B}(\Pi)]\vee\mathcal{N}, \mathcal{B}(\mathbb{R}))$ be bounded, Φ a co-moderate Young function and $\Psi(t) = t \log^+ t$, $t \geq 0$. Then

$$\| \sup_{t\in\Pi} | {}^{\alpha_i}X_t | \|_\Phi \leq 2\frac{p}{p-1} \| \sup_{t\in\Pi} |X_t| \|_\Phi \quad , \alpha=\gamma,\pi.$$

In particular, for $q>1$

$$\| \sup_{t\in\Pi} | {}^{\alpha_i}X_t | \|_q \leq 2\frac{q}{q-1} \| \sup_{t\in\Pi} |X_t| \|_q \quad , \alpha=\gamma,\pi.$$

Moreover, there exists a constant c which does not depend on X such that

$$\| \sup_{t\in\Pi} | {}^{\alpha_i}X_t | \|_1 \leq c \| \sup_{t\in\Pi} |X_t| \|_\Psi \quad , \alpha=\gamma,\pi.$$

<u>Proof</u>:

Let $\alpha=\gamma,\pi$, Φ', Ψ' the conjugate functions of Φ,Ψ. Compare Dellacherie, Meyer [18], pp. 178-186 for this and the following arguments. By an easy computation

$$\Psi'(t) = t\, 1_{[0,1]}(t) + \exp(t-1)\, 1_{]1,\infty[}(t), \quad t \geq 0.$$

To abbreviate, let $V^* = \sup_{t\in\Pi} |V_t|$ for a process V, set $Y = {}^{\alpha_i}X$ and

$Z = {}^{\alpha_i}X^*$. Clearly, $X \le X^*$ implies $|Y| \le Z$. Therefore we have to prove that there exists a constant c independent of X such that

$$\| Z^* \|_\Phi \le 2\frac{p}{p-1} \| X^* \|_\Phi \quad ,$$

$$\| Z^* \|_1 \le c \| X^* \|_\Psi \ .$$

Denote by $\mathbb{A}$ the set of all integrable increasing processes such that $\| A_{(1,1)} \|_{\Phi'} \le 1$ respectively $A_{(1,1)} \le 1$. By a classical section theorem (see Dellacherie, Meyer [17], pp. 103-105) followed by theorems 1 and 2 we have

$$\begin{aligned} \| Z^* \|_\Phi &\le \sup_{A\in\mathbb{A}} \int_{\Omega\times\Pi} Z \, dm_A = \sup_{A\in\mathbb{A}} \int_{\Omega\times\Pi} X^* \, dm_{A^{\alpha_i}} \\ &= \sup_{A\in\mathbb{A}} E(X^* A^{\alpha_i}_{(1,1)}) \end{aligned}$$

and an analogous inequality for $\| Z^* \|_1$. Young's inequality now yields

$$\begin{aligned} \sup_{A\in\mathbb{A}} E(X^* A^{\alpha_i}_{(1,1)}) &\le 2 \| X^* \|_\Phi \sup_{A\in\mathbb{A}} \| A^{\alpha_i}_{(1,1)} \|_{\Phi'} \quad \text{respectively} \\ &\le 2 \| X^* \|_\Psi \sup_{A\in\mathbb{A}} \| A^{\alpha_i}_{(1,1)} \|_{\Psi'} . \end{aligned}$$

In the first case we can conclude using theorem 3. For the second case, observe that if $A_{(1,1)} \le 1$ for $A\in\mathbb{A}$, theorem 3 yields a constant $c>0$ such that

$$\sup_{A\in\mathbb{A}} E(\exp(A^{\alpha_i}_{(1,1)}/c))<\infty .$$

This implies the desired inequality in the second case. □

5. One-directional decomposition theorems

One aspect of the theorem on the existence of dual previsible projections has not yet been emphasized. According to theorem (4.2) for any integrable increasing process A there exists exactly one i-previsible integrable increasing process A^{π_i} such that $m_A|\mathcal{P}^i = m_{A^{\pi_i}}|\mathcal{P}^i$. In other words: there is exactly one i-previsible integrable increasing process A^{π_i} such that the Doléans measure of $A-A^{\pi_i}$ is zero or such that $A-A^{\pi_i}$ is an i-martingale, i=1,2. But this means that theorem (4.2) is the simplest form of a one-directional decomposition theorem.

More generally, decomposition theorems deal with the representation of submartingales by martingales and increasing processes. The above remarks contain the key observation for the decomposition method applied here. Let X be an i-submartingale. We first have to make sure that the Doléans measure m_X exists. If this is granted, we simply take the dual i-previsible projection of the increasing process associated with m_X to decompose X into an i-martingale and an i-previsible increasing process. We also treat weak submartingales whose "marginal" processes are one-parameter submartingales in this way and give explicit approximations in the L^1-weak sense for their increasing processes. Inequalities of the Garsia-Neveu type, comparing norms of submartingales with norms of their increasing processes conclude this section. Our presentation follows Brennan [5], Dozzi [20], and Métivier [31].

The following property will prove to be characteristic for the existence of Doléans measures of i-submartingales.

Definition 1: Let X be a process such that X_t is integrable for any $t \in \mathbb{I}$. X is said to be "of class D^1", if for any 0-sequence $(\mathbb{J}_n)_{n \in \mathbb{N}}$ of partitions of [0,1] the family

$$\{ \sum_{J \in \mathbb{J}_n} E(\Delta_J X_{(\cdot,1)} | \mathcal{F}^1_{sJ}) \ : \ n \in \mathbb{N}\}$$

is uniformly integrable. Analogously, the notion "of class D^2" is defined.

Proposition 1: Let i=1,2, X an i-submartingale which is $\hat{Q}^{++}$-continuous in $L^1(\Omega,\mathcal{F},P)$. Then m_X exists iff X is of class D^i.

Proof:

Our arguments are for i=1.

1. Assume X is of class D^1. Fix a 0-sequence $(\mathbb{I}_n)_{n\in\mathbb{N}}$ of partitions of [0,1] and for $t\in\mathbb{I}$ set

$$L_t = \{\sum_{J\in\mathbb{I}_n} E(\Delta_{J\cap[0,t_1]\times[0,t_2]}X|\mathcal{F}^1_{sJ}) \; : \; n\in\mathbb{N}\}.$$

By the compactness criterion of Dunford and Pettis (see for example Dellacherie, Meyer [17], p. 43) and the 1-submartingale property of X each L_t is weakly relatively compact. We can therefore choose a subsequence of $(\mathbb{I}_n)_{n\in\mathbb{N}}$ such that the corresponding subsequences of L_t converge weakly for all $t\in\mathbb{I}\cap\mathbb{Q}^2$ to some nonnegative random variable V_t. We may suppose that $(V_t)_{t\in\mathbb{I}\cap\mathbb{Q}^2}$ is increasing and therefore define

$$A_t = \lim_{\substack{\mathbb{Q}^2\ni s\to t\\ s\geq t}} V_s, \; t\in\mathbb{I}.$$

Then A is an integrable increasing process. Moreover, since X is $\hat{Q}^{++}$-continuous in $L^1(\Omega,\mathcal{F},P)$, the equation

$$E(1_F\, \Delta_J A) = E(1_F\, \Delta_J X) \;, \; F\times J\in\mathcal{R}^1,$$

holds. This entails $m^o_X = m^o_A$ and $m_A|\mathcal{P}^1$ is the extension we are looking for.

2. Assume m_X exists. Again, fix a 0-sequence $(\mathbb{I}_n)_{n\in\mathbb{N}}$ of partitions of [0,1]. Then, in the above notations

$$\sup_{\xi\in L_1} \|\xi\|_1 = m_X(\Omega\times\mathbb{I}) < \infty.$$

Let A be the integrable increasing process associated with m_X according to proposition (2.3). Then for arbitrary $F\in\mathcal{F}$, $n\in\mathbb{N}$

$$\begin{aligned} E(1_F \sum_{J\in\mathbb{I}_n} E(\Delta_J X_{(\cdot,1)}|\mathcal{F}^1_{sJ})) &= E(\sum_{J\in\mathbb{I}_n} E(1_F|\mathcal{F}^1_{sJ})\; \Delta_J X_{(\cdot,1)}) \\ &= E(\sum_{J\in\mathbb{I}_n} E(1_F|\mathcal{F}^1_{sJ})\; \Delta_J A_{(\cdot,1)}) \\ &\leq E(\sup_{t\in\mathbb{I}} |{}^{\gamma_1}(1_F)_t|\; A_{(1,1)}). \end{aligned}$$

Since $1_F \leq 1$, also ${}^{\gamma_1}(1_F) \leq 1$. Therefore, using theorem (4.4), we obtain

$$\lim_{P(F)\to 0} E(\sup_{t\in \Pi} |{}^{\gamma_1}(1_F)_t| A_{(1,1)}) = 0.$$

Summing up, we have established a well known criterion for uniform integrability of L_1. □

The following decomposition theorem for i-submartingales of class D^i is guaranteed by proposition 1.

Theorem 1: Let X be a 1-submartingale of class D^1 which is $\hat{Q}^{++}$-continuous in $L^1(\Omega,\mathcal{F},P)$. There exists a unique integrable increasing process $A \in \mathfrak{M}(\mathcal{P}^1, \mathcal{B}(\mathbb{R}))$ such that X-A is a 1-martingale, or such that $m_X|\mathcal{P}^1 = m_A|\mathcal{P}^1$. Moreover, for any 0-sequence $(\mathbb{I}_n)_{n\in\mathbb{N}}$ of partitions of [0,1], any $t\in \Pi$

$$A_t = \lim_{n\to\infty} \sum_{J\in \mathbb{I}_n} E(\Delta_{J\cap[0,t_1]\times[0,t_2]}X|\mathcal{F}^1_{sJ}) \text{ weakly in } L^1(\Omega,\mathcal{F},P).$$

A corresponding statement holds for the second parameter.

Proof:

The Doléans measure m_X of X exists by proposition 1. Let A be the dual 1-previsible projection of the integrable increasing process associated with m_X. Then $m_X|\mathcal{P}^1 = m_A|\mathcal{P}^1$ according to theorem (4.2). Hence A is appropriate and existence is established. Uniqueness follows again from theorem (4.2).

Now let a 0-sequence $(\mathbb{I}_n)_{n\in\mathbb{N}}$ of partitions of [0,1] be given. For simplicity, let t=(1,1). Fix an arbitrary $F\in\mathcal{F}$ and denote by M a right-continuous version of the one-parameter martingale $E(1_F|\mathcal{F}^1_\cdot)$ possessing left limits. Then for any $n\in\mathbb{N}$ like in part 2 of the proof of proposition 1

$$E(1_F \sum_{J\in \mathbb{I}_n} E(\Delta_J X_{(\cdot,1)}|\mathcal{F}^1_{sJ}) = E(\sum_{J\in \mathbb{I}_n} E(1_F|\mathcal{F}^1_{sJ})\, \Delta_J A_{(\cdot,1)})$$

and therefore

$$\lim_{n\to\infty} E(1_F \sum_{J\in \mathbb{I}_n} E(\Delta_J X_{(\cdot,1)}|\mathcal{F}^1_{sJ})) = \int_{\Omega\times\Pi} M^{-\cdot}\, dm_A = E(1_F\, A_{(1,1)}).$$

For the last equality sign we have used remark 2 following definition (4.1) and theorem (4.2). Since F is arbitrary, the assertion follows. □

Definition 2: Let $i=1,2$, X an i-submartingale of class D^i which is $\hat{Q}^{++}$-continuous in $L^1(\Omega,\mathcal{F},P)$. The integrable increasing process A according to theorem 1 is called "dual i-previsible projection" of X.

We go one more step ahead and consider decomposition theorems for weak submartingales. The weak submartingale property alone, however, will not be enough. In addition, we require a submartingale property in the order sense. The prototype of processes satisfying these requirements is the square of a square integrable martingale (see prop. (9.1)). At first, we have to verify a property which, in the case of i-submartingales, was trivial.

Proposition 2: Let $i=1,2$, X a weak submartingale such that $X_{(.,t_{\bar{i}})}$ is an $\mathbb{F}_i$-submartingale for any $t_{\bar{i}}\in[0,1]$. If X is of class D^i, then $X_{(.,t_{\bar{i}})}$ is of class D^i for all $t_{\bar{i}}\in[0,1]$.

Proof:

Assume $i=1$, pick $t_2\in[0,1]$ and let $(\mathbb{J}_n)_{n\in\mathbb{N}}$ be a 0-sequence of partitions of [0,1]. Use the uniform integrability criterion of de la Vallée-Poussin (see Dellacherie, Meyer [17], p. 38) to choose a Young function Φ such that

$$\sup_{n\in\mathbb{N}} \| \sum_{J\in\mathbb{J}_n} E(\Delta_J X_{(.,1)}|\mathcal{F}^1_{sJ}) \|_{\Phi} < \infty.$$

Remember Φ is increasing and convex. The weak submartingale property of X and Jensen's inequality therefore imply for $n\in\mathbb{N}$

$$\| \sum_{J\in\mathbb{J}_n} E(\Delta_J X_{(.,t_2)}|\mathcal{F}^1_{sJ}) \|_{\Phi} \leq \| \sum_{J\in\mathbb{J}_n} E(E(\Delta_J X_{(.,1)}|\mathcal{F}^1_{sJ})|\mathcal{F}^2_{t_2}) \|_{\Phi}$$
$$\leq \| \sum_{J\in\mathbb{J}_n} E(\Delta_J X_{(.,1)}|\mathcal{F}^1_{sJ}) \|_{\Phi} .$$

Now apply de la Vallée-Poussin's criterion in the reverse direction. □

Theorem 2: Let X be a weak submartingale of class D^1 which is $\hat{Q}^{++}$-continuous in $L^1(\Omega,\mathcal{F},P)$ such that $X_{(.,t_2)}$ is an $\mathbb{F}_1$-submartingale for any $t_2 \in [0,1]$. Then there exists a unique 2-submartingale A such that

i) $A_{(.,t_2)}$ is the dual previsible projection of $X_{(.,t_2)}$ for all $t_2 \in [0,1]$,

ii) X-A is a 1-martingale.

Moreover, for any 0-sequence $(\mathbb{J}_n)_{n\in\mathbb{N}}$ of partitions of $[0,1]$, any $t \in \mathbb{I}$

$$A_t = \lim_{n\to\infty} \sum_{J\in\mathbb{J}_n} E(\Delta_{J\cap[0,t_1]\times[0,t_2]}X|\mathcal{F}^1_{s^J}) \text{ weakly in } L^1(\Omega,\mathcal{F},P).$$

A corresponding statement holds for the second parameter.

Proof:

According to proposition 2 and theorem 1, $X_{(.,t_2)}$ possesses a dual 1-previsible projection $A_{(.,t_2)}$ for all $t_2 \in [0,1]$. The process A satisfies i) and ii) by definition and the representation equation by theorem 1. It remains to show that A is a 2-submartingale. To this end, let $K \in J$ be given and choose a 0-sequence $(\mathbb{J}_n)_{n\in\mathbb{N}}$ of partitions of $[0,1]$ which is finer than the partition defined by K_1. Then in consequence of the weak submartingale property of X

$$E(\Delta_K A|\mathcal{F}^2_{s_2^K}) = \lim_{n\to\infty} \sum_{\mathbb{J}_n \ni J\subset K_1} E(\Delta_{J\times K_1}X|\mathcal{F}_{(s^J,s_2^K)}) \geq 0.$$

This completes the proof. □

Remark: Unlike theorem 1, theorem 2 does not contain a statement on the $\mathcal{P}^i$-measurability of the process A appearing in the decomposition of the given weak submartingale. We were unable to verify whether $A \in \mathbb{M}(\mathcal{P}^i, \mathcal{B}(\mathbb{R}))$. Compare, however, Stricker, Yor [40], pp. 116-117, where a condition is discussed under which it is true.

<u>Definition 3</u>: Let $i=1,2$, X a weak submartingale of class D^i which is $\hat{Q}^{++}$-continuous in $L^1(\Omega,\mathcal{F},P)$ such that $X_{(\cdot,t_{\bar{i}})}$ is an $\mathbb{F}_i$-submartingale for any $t_{\bar{i}} \in [0,1]$. The unique $\bar{i}$-submartingale A according to theorem 2 is called "<u>dual i-previsible projection</u>" of X.

We finally extend the inequalities of theorem (4.3) to the dual previsible projections constructed in theorems 1 and 2. The one-parameter counterparts of the inequalities we consider are attributed to Garsia and Neveu (see Dellacherie, Meyer [18], p. 183).

<u>Theorem 3</u>: Let $i=1,2$, Φ a moderate, Ψ a moderate and co-moderate Young function, p and q their respective exponents.

1. If $X \in \mathfrak{M}([\mathcal{F}\times\mathcal{B}(\Pi)]\vee\mathcal{N}, \mathcal{B}(\mathbb{R}))$ is an i-submartingale of class D^i which is $\hat{Q}^{++}$-continuous in $L^1(\Omega,\mathcal{F},P)$, A its dual i-previsible projection, we have

$$\| A_{(1,1)} \|_{\Phi} \leq 2p \, \| \sup_{t_i\in[0,1]} |X_{(t_i,1)}| \, \|_{\Phi}$$

and in particular for $r \geq 1$

$$\| A_{(1,1)} \|_{r} \leq 2r \, \| \sup_{t_i\in[0,1]} |X_{(t_i,1)}| \, \|_{r} .$$

2. If $X \in \mathfrak{M}([\mathcal{F}\times\mathcal{B}(\Pi)]\vee\mathcal{N}, \mathcal{B}(\mathbb{R}))$ is a weak submartingale of class D^i which is $\hat{Q}^{++}$-continuous in $L^1(\Omega,\mathcal{F},P)$ such that $X_{(\cdot,t_{\bar{i}})}$ is an $\mathbb{F}_i$-submartingale for each $t_{\bar{i}}\in[0,1]$, A a version of its dual i-previsible projection for which $A_{(1,\cdot)}$ is right-continuous, we have

$$\| \sup_{t_{\bar{i}}\in[0,1]} A_{(1,t_{\bar{i}})} \|_{\Psi} \leq 2\, q^2/(q-1) \; \| \sup_{t_i\in[0,1]} |X_{(t_i,1)}| \, \|_{\Psi}$$

and in particular for $r>1$

$$\| \sup_{t_{\bar{i}}\in[0,1]} A_{(1,t_{\bar{i}})} \|_{r} \leq 2r^2/(r-1) \, \| \sup_{t_i\in[0,1]} |X_{(t_i,1)}| \, \|_{r} .$$

If X is nonnegative, the factor 2 on the right hand side of the inequalities in both parts can be replaced by 1.

Proof:

Assume i=1.

1. Observe that $X_{(.,1)}$ is a one-parameter submartingale whose dual previsible projection is $A_{(.,1)}$. The potential

$$Z = E(A_{(1,1)}-A_{(.,1)}|\mathcal{F}^1_.) = E(X_{(1,1)}-X_{(.,1)}|\mathcal{F}^1_.)$$

is majorized by the martingale $2\ E(\sup_{t_i\in[0,1]} |X_{(t_1,1)}|\ |\mathcal{F}^1_.)$. In case X is nonnegative, the factor 2 can be replaced by 1. Therefore the first inequality is a consequence of the inequality of Garsia-Neveu (see Dellacherie, Meyer [18], p. 181).

2. Note that $A_{(1,.)}$ is a right-continuous nonnegative submartingale. Therefore, the maximal inequality of Doob (see Dellacherie, Meyer [18], p. 72 or p. 184) yields

$$\|\sup_{t_2\in[0,1]} A_{(1,t_2)}\|_\Psi \leq q/(q-1)\ \ \|A_{(1,1)}\|_\Psi\ .$$

The right hand side of this inequality can now be treated like in part 1. □

6. Regularity of one-directional projections

In the theory of one-parameter processes, regularity properties are nicely transferred from processes to their projections (see Dellacherie, Meyer [18], pp. 119-123). We will consider this problem now for one-directional projections. For half-planes, the results are as follows: if X possesses limits in $Q^{+\pm}$, $Q^{-\pm}$, then so do ${}^{\gamma_1}X$, ${}^{\pi_1}X$; if X is $Q^{+\pm}$-continuous, then so is ${}^{\gamma_1}X$; if X is $Q^{-\pm}$-continuous, ${}^{\pi_1}X$ has the same property. For the quadrants, things turn out to be more difficult: ${}^{\gamma_1}X$, ${}^{\pi_1}X$ inherit from X the property of possessing Q^{++}-, Q^{+-}-limits and if X is

Q^{++}-continuous, so is ${}^{\gamma_1}X$. But a counterexample shows that for the left hand quadrants Q^{-+} and Q^{--} nothing can be said in general. Of course, similar results hold for projections in direction 2.

Since a martingale M is just the iterated 1- and 2-optional projection of $M_{(1,1)}$ (see section 8), we can and will use the results of this section for deriving regularity properties of martingales. The above mentioned anomaly indicates that for the existence of Q^{--}-limits, we will need more than we can cover here (see section 9). The material presented in this section was published in a paper of Millet, Sucheston [34]. Bakry [3] translated the "amart" techniques used there into the language of the "general theory" of two-parameter processes we employ.

We start with two auxiliary propositions in which the problem of existence of limits is treated at stopping points. The first one is for right hand limits.

<u>Proposition 1</u>: Let $X \in \mathfrak{M}([\mathcal{F}\times\mathcal{B}(\Pi)]\vee\mathcal{N}, \mathcal{B}(\mathbb{R}))$ be bounded, T a 1-stopping point. Then:

1. if $\underline{X}_T^{+\pm} = \overline{X}_T^{+\pm}$, the same is true for ${}^{\gamma_1}X$, ${}^{\pi_1}X$,
2. if $\underline{X}_T^{++} = \overline{X}_T^{++}$ resp. $\underline{X}_T^{+-} = \overline{X}_T^{+-}$, the same is true for ${}^{\gamma_1}X$, ${}^{\pi_1}X$,
3. if $X_T = \underline{X}_T^{+\pm} = \overline{X}_T^{+\pm}$, the same is true for ${}^{\gamma_1}X$,
4. if $X_T = \underline{X}_T^{++} = \overline{X}_T^{++}$, the same is true for ${}^{\gamma_1}X$.

An analogous statement holds for the second parameter.

<u>Proof</u>:

To abbreviate, set $Y = {}^{\alpha_1}X$ for $\alpha=\gamma,\pi$. For simplicity, let $|X|\le 1$. Assume

$$B^{\alpha} = \{\underline{Y}_T^{+\pm} < \overline{Y}_T^{+\pm}\},$$

which is measurable by theorem (2.1), has positive probability. Pick $a,b\in\mathbb{R}$, $a<b$, such that

$$C^{\alpha} = \{\underline{Y}_T^{+\pm} \le a,\ b \le \overline{Y}_T^{+\pm}\}$$

still has positive probability. Use proposition (3.3) to choose a sequence $(R_n)_{n\in\mathbb{N}}$ of 1-previsible sets such that

(6.1) $[(R_n)_\omega]^a$ is a neighborhood of T_ω in $[Q^{+\pm}_{T_\omega}]^a$ for $\omega\in\Omega$, $n\in\mathbb{N}$,

(6.2) $\operatorname{diam}(R_n) \underset{n\to\infty}{\to} 0.$

For $n\in\mathbb{N}$, define

$$\underline{C}^\alpha_n = \{Y\leq a\}\cap R_n, \qquad \overline{C}^\alpha_n = \{Y\geq b\}\cap R_n.$$

We know that $\underline{C}^\alpha_n, \overline{C}^\alpha_n \in \mathfrak{G}^1$ resp. $\mathfrak{P}^1$, $C^\alpha = \pi_\Omega(\underline{C}^\alpha_n) \cap \pi_\Omega(\overline{C}^\alpha_n)$, $\alpha=\gamma,\pi$, $n\in\mathbb{N}$, by (6.1). Apply theorem (3.1) to find (previsible) 1-stopping points $\underline{T}^\alpha_n$, $\overline{T}^\alpha_n$ such that $\underline{T}^\alpha_n = (1,1)$ on $\{\underline{T}^\alpha_n \notin \underline{C}^\alpha_n\}$, $P(\pi_\Omega(\underline{C}^\alpha_n)\cap\{\underline{T}^\alpha_n\notin\underline{C}^\alpha_n\}) < 2^{-n}$ and a corresponding statement for $\overline{T}^\alpha_n$, $\overline{C}^\alpha_n$.

Consider the integrable increasing processes

$$\underline{A}^\alpha_n = 1_{\pi_\Omega(\underline{C}^\alpha_n)}\, 1_{[\underline{T}^\alpha_n,(1,1)]}, \qquad \overline{A}^\alpha_n = 1_{\pi_\Omega(\overline{C}^\alpha_n)}\, 1_{[\overline{T}^\alpha_n,(1,1)]}$$

which are in $\mathfrak{M}(\mathfrak{G}^1, \mathcal{B}(\mathbb{R}))$ resp. $\mathfrak{M}(\mathfrak{P}^1, \mathcal{B}(\mathbb{R}))$ if $\alpha=\gamma$ resp. $\alpha=\pi$, $n\in\mathbb{N}$. By theorem (4.1) for all $n\in\mathbb{N}$

$$b\,P(C^\alpha) - 2^{-n} \leq E(1_{\pi_\Omega(\overline{C}^\alpha_n)} Y_{\overline{T}^\alpha_n}) = \int_{\Omega\times\Pi} Y\, d\overline{A}^\alpha_n$$

$$= \int_{\Omega\times\Pi} X\, d\overline{A}^\alpha_n = E(1_{\pi_\Omega(\overline{C}^\alpha_n)} X_{\overline{T}^\alpha_n})$$

and similarly

$$a\,P(C^\alpha) + 2^{-n} \geq E(1_{\pi_\Omega(\underline{C}^\alpha_n)} X_{\underline{T}^\alpha_n}).$$

Now if $\underline{X}^{+\pm}_T = \overline{X}^{+\pm}_T$, (6.2) allows to conclude

$$a\,P(C^\alpha) \geq b\,P(C^\alpha),$$

which contradicts $a<b$. Assertion 1 follows.

To prove assertion 3, consider

$$B^\alpha_1 = \{Y_T < \overline{Y}^{+\pm}_T\}, \qquad B^\alpha_2 = \{\underline{Y}^{+\pm}_T < Y_T\}$$

instead of C^α and conclude, using $X_T = \overline{X}^{+\pm}_T$, $\underline{X}^{+\pm}_T = X_T$.

The arguments for the second and fourth, finally, are almost identical to the ones presented and need an appeal to proposition (3.4) instead of (3.3). □

For the left hand limits we can only prove a "half plane" version.

Proposition 2: Let $X \in \mathfrak{M}([\mathcal{F}\times\mathcal{B}(\mathbb{I})]\vee\mathcal{N}, \mathcal{B}(\mathbb{R}))$ be bounded, T a previsible 1-stopping point. Then:

1. if $\underline{X}_T^{-\pm} = \overline{X}_T^{-\pm}$, the same is true for ${}^{\gamma}{}_1X$, ${}^{\pi}{}_1X$,
2. if $X_T = \underline{X}_T^{-\pm} = \overline{X}_T^{-\pm}$, the same is true for ${}^{\pi}{}_1X$.

An analogous statement holds for the second parameter.

Proof:

The proof of proposition 1 works with only one modification. The second part of proposition (3.3) is needed. □

The stage is set for the derivation of regularity properties of one-directional projections. We start with half plane limits.

Theorem 1: Let $X \in \mathfrak{M}([\mathcal{F}\times\mathcal{B}(\mathbb{I})]\vee\mathcal{N}, \mathcal{B}(\mathbb{R}))$ be bounded. Then:

1. if X possesses $Q^{+\pm}$ resp. $Q^{-\pm}$-limits, so does ${}^{\gamma}{}_1X$ and ${}^{\pi}{}_1X$,
2. if X is $Q^{+\pm}$-continuous, so is ${}^{\gamma}{}_1X$,
3. if X is $Q^{-\pm}$-continuous, so is ${}^{\pi}{}_1X$.

An analogous statement holds for the second parameter.

Proof:

1. Let $Y = {}^{\alpha}{}_1X$, $\alpha=\gamma,\pi$. By proposition (2.1) and theorem (2.1),

$$S^+ = \{\underline{Y}^{+\pm} < \overline{Y}^{+\pm}\} \in \mathfrak{M}^1, \qquad S^- = \{\underline{Y}^{-\pm} < \overline{Y}^{-\pm}\} \in \mathfrak{P}^1.$$

Assume S^+ or S^- is non-evanescent. Then we can find numbers $a,b \in \mathbb{R}$, $a<b$, such that

$$V^+ = \{\underline{Y}^{+\pm} \leq a,\ b \leq \overline{Y}^{+\pm}\} \quad \text{or} \quad V^- = \{\underline{Y}^{-\pm} \leq a,\ b \leq \overline{Y}^{-\pm}\}$$

is still non-evanescent. If V^+ is non-evanescent, we can apply proposition (3.1) to find a 1-stopping point T^+ such that

$$P(\underline{Y}^{+\pm}_{T^+} \leq a,\ b \leq \overline{Y}^{+\pm}_{T^+}) > 0.$$

If V^- is non-evanescent, theorem (3.1) produces a previsible 1-stopping point T^- such that

$$P(\underline{Y}^{-\pm}_{T^-} \leq a,\ b \leq \overline{Y}^{-\pm}_{T^-}) > 0.$$

The first parts of propositions 1 and 2 allow to conclude.

2. Assume X is $Q^{+\pm}$-continuous. Then $Y = {}^{\gamma}1X$ possesses $Q^{+\pm}$-limits according to 1. and

$$Y^{+\pm} = \lim_{n\to\infty} \sum_{1\le k\le n} Y^{+\pm}_{(.,k/n)}\, 1_{[(k-1)/n,k/n[} \in \mathfrak{m}(\mathfrak{G}^1, \mathcal{B}(\mathbb{R})),$$ since

$\mathbb{F}_1$ is right-continuous. Therefore

$$W = \{Y \neq Y^{+\pm}\} \in \mathfrak{G}^1.$$

Assume W is non-evanescent. By theorem (3.1) there exists a 1-stopping point U such that

$$P(Y_U \neq Y_U^{+\pm}) > 0.$$

Part 3 of proposition 1 yields the conclusion.

Finally, if X is $Q^{-\pm}$-continuous, a similar argument using part 2 of proposition 2 gives the last part of the assertion. □

We next consider quadrant limits.

<u>Theorem 2</u>: Let $X \in \mathfrak{m}([\mathcal{F}\times\mathcal{B}(\Pi)]\vee\mathcal{N}, \mathcal{B}(\mathbb{R}))$ be bounded. Then:

1. if X possesses Q^{++}-limits and Q^{+-}-limits, so does ${}^{\gamma}1X$ and ${}^{\pi}1X$,
2. if X is Q^{++}-continuous, so is ${}^{\gamma}1X$.

An analogous statement holds for the second parameter.

<u>Proof</u>:

We have to refine the proof of theorem 1. Let again $Y = {}^{\alpha}1X$, $\alpha=\gamma,\pi$.
By proposition (2.1) and theorem (2.1), $\underline{Y}^{++}$, $\underline{Y}^{+-}$, $\overline{Y}^{++}$, $\overline{Y}^{+-}$ are $\mathfrak{m}^1$-measurable. Hence

$$S^+ = \{\underline{Y}^{++} < \overline{Y}^{++}\} \in \mathfrak{m}^1, \quad S^- = \{\underline{Y}^{+-} < \overline{Y}^{+-}\} \in \mathfrak{m}^1.$$

Assume S^+ or S^- is non-evanescent and choose $a,b \in \mathbb{R}$, $a<b$, such that for

$$U^+ = \{\underline{Y}^{++} \le a,\ b \le \overline{Y}^{++}\} \quad \text{or} \quad U^- = \{\underline{Y}^{+-} \le a,\ b \le \overline{Y}^{+-}\}$$

this is still the case. Now apply proposition (3.1) to $U^+ \cup U^-$ to find a 1-stopping point T such that

$$P(\{\underline{Y}_T^{++} \le a,\ b \le \overline{Y}_T^{++}\} \cup \{\underline{Y}_T^{+-} \le a,\ b \le \overline{Y}_T^{+-}\}) > 0.$$

Part 2 of proposition 1 yields the first assertion.

Now assume X is Q^{++}-continuous and let $\alpha=\gamma$. From part 1 we know that Y has Q^{++}-limits and therefore

$$Y^{++} = \lim_{n\to\infty} \sum_{1\le k\le n} Y^{++}_{(.,k/n)} 1_{[(k-1)/n,k/n[} \in \mathbb{M}(G^1, \mathcal{B}(\mathbb{R})).$$

From here on we can continue along the lines of reasoning of the second part of the proof of theorem 1. Instead of part 3 of proposition 1 we have to apply part 4. □

By adapting an example of Bakry [4], we show now that theorems 1 and 2 is all we can get. We construct a process X which has limits in all four quadrants, is Q^{++}-continuous and whose 1-projections fail to have Q^{-+}- and Q^{--}-limits.

Example:

Let $\Omega = [0,1[$, $\mathcal{F} = \mathcal{B}(\Omega)$, $P = \lambda|\mathcal{F}$. Moreover, let $(r_n)_{n\in\mathbb{N}}$ be a strictly increasing sequence in $[0,1]$ such that $r_1=0$, $r_n\uparrow 1$. For $n\in\mathbb{N}$ let $\mathcal{B}_n$ be the algebra generated by the dyadic intervals $J^{i,n} = \{[(i-1)/2^n, i/2^n[: 1\le i\le 2^n\}$. For $t_1\in[r_n,r_{n+1}[$, take $\mathcal{F}^1_{t_1}$ to be $\mathcal{B}_n$, completed by the 0-sets of $\mathcal{F}$, and for all $t_2\in[0,1]$, set $\mathcal{F}^2_{t_2} = \mathcal{F}$. The filtration $\mathcal{F}_t = \mathcal{F}^1_{t_1}\cap\mathcal{F}^2_{t_2}$, $t\in\Pi$, is right-continuous and clearly possesses the property of conditional independence. We consider the process

$$X_t(\omega) = 1_{[0,t_2]}(\omega), \quad \omega\in\Omega, \ t\in\Pi.$$

Obviously X is $Q^{\pm+}$-continuous and possesses $Q^{\pm-}$-limits. In particular, X is Q^{++}-continuous and possesses Q^{+-}-, Q^{-+}-, and Q^{--}-limits. Now for $t\in\Pi$ let

$$Y_t = \sum_{n\in\mathbb{N}} 1_{[r_n,r_{n+1}[}(t_1) \Big[\sum_{1\le i\le[2^n t_2]} 1_{J^{i,n}} + (2^n t_2-[2^n t_2])\, 1_{J^{[2^n t_2]+1,n}}\Big] + 1_{\{1\}}(t_1)\, 1_{[0,t_2]},$$

$$Z_t = \sum_{n\in\mathbb{N}} 1_{]r_n,r_{n+1}]}(t_1) \Big[\sum_{1\le i\le[2^n t_2]} 1_{J^{i,n}} + (2^n t_2-[2^n t_2])\, 1_{J^{[2^n t_2]+1,n}}\Big]$$

$$+ 1_{\{1\}}(t_1)\, 1_{[0,t_2]}.$$

Moreover, for $n \in \mathbb{N}$ let

$$Y^n = 1_{\Omega\times([0,r_{n+1}[\cup\{1\})\times[0,1]}\, Y.$$

Then Y^n is regular and adapted and $Y^n \to Y$. Hence Y is optional, in particular 1-optional. In the same way, it is seen that Z is 1-previsible. Finally,

$$Y_t = E(X_t|\mathcal{F}^1_{t_1}) \quad \text{for } t\in \Pi.$$

Therefore, as follows from theorem (4.1), $Y = {}^{Y_1}X$, $Z = {}^{\pi_1}X$. We will prove that neither Y nor Z possess Q^{-+}- or Q^{--}-limits. Since the arguments for Y and Z are almost identical, we concentrate on Y. We fix an arbitrary non-dyadic $\omega\in\Omega$ and show

$$(6.3)\quad \underline{Y}^{-+}(\omega) \neq \overline{Y}^{-+}(\omega), \quad \underline{Y}^{--}(\omega) \neq \overline{Y}^{--}(\omega).$$

For each $n\in\mathbb{N}$, there is exactly one dyadic interval $[a_n,b_n[$ of length 2^{-n} which contains ω. Let

$$s^n_2 = a_n + \tfrac{1}{4}2^{-n}, \quad t^n_2 = a_n + \tfrac{1}{2}2^{-n}, \quad u^n_2 = a_n + \tfrac{3}{4}2^{-n},$$
$$s^n_1 = t^n_1 = u^n_1 = r_n, \; n\in\mathbb{N}.$$

Since ω is non-dyadic, there are infinitely many $k\in\mathbb{N}$ such that s^k_2, $t^k_2 < \omega$ and infinitely many $l\in\mathbb{N}$ such that t^l_2, $u^l_2 > \omega$. Hence we can choose subsequences $(v^n)_{n\in\mathbb{N}}$ of $(s^n)_{n\in\mathbb{N}}$, $(w^n)_{n\in\mathbb{N}}$ of $(t^n)_{n\in\mathbb{N}}$, $(x^n)_{n\in\mathbb{N}}$ of $(t^n)_{n\in\mathbb{N}}$ and $(y^n)_{n\in\mathbb{N}}$ of $(u^n)_{n\in\mathbb{N}}$ which converge to $(1,\omega)$ such that

$$v^n,\, w^n \in Q^{--}_{(1,\omega)}, \quad Y_{v^n}(\omega) = \tfrac{1}{4}, \quad Y_{w^n}(\omega) = \tfrac{1}{2},$$
$$x^n,\, y^n \in Q^{-+}_{(1,\omega)}, \quad Y_{x^n}(\omega) = \tfrac{1}{2}, \quad Y_{y^n}(\omega) = \tfrac{3}{4}$$

for all $n\in\mathbb{N}$. Consequently

$$\underline{Y}^{-+}_{(1,\omega)}(\omega) \neq \overline{Y}^{-+}_{(1,\omega)}(\omega), \quad \underline{Y}^{--}_{(1,\omega)}(\omega) \neq \overline{Y}^{--}_{(1,\omega)}(\omega).$$

This implies (6.3) which was proposed to be shown.

II. Two-parameter processes

Although we encountered the property of conditional independence of $\mathbb{F}_1$ and $\mathbb{F}_2$ at a few places in chapter I, it is in this chapter that its full impact on the theory will be felt. In terms of projections, it essentially says: 1- and 2-projections are independent of each other and commute. As a consequence, the optional (previsible) sets turn out to be just the intersection of the 1- and 2-optional (previsible) sets and optional (previsible) projections can be defined as the product of two one-directional ones in an arbitrary order. However, the proof of these facts is not easy. It rests upon a regularity theorem for the most elementary two-directional projections, the martingales. Though, due to the poorness of the concept of stopping points, the proof of the existence of regular versions cannot be given unless a genuinely two-directional stopping notion is studied. We investigate "stopping lines" in section 7. Unlike stopping points, they are able to define a "past" and a "future" in a two-parameter setting. Sections 8 and 9 are devoted to the proof of the regularity theorem for martingales. In the preparatory section 8 we derive the main martingale inequalities from their one-parameter counterparts: the inequalities of Doob - Cairoli and of Burkholder - Davis. With their help we are able, at the beginning of section 9, to show that for square integrable martingales M, the submartingale M^2 possesses a Doléans measure. That makes it possible to define the stochastic integral of M and study the corresponding integral process. This process tells us what it means to "stop a martingale before a previsible stopping line", a concept which is needed in the proof of the existence of

Q^{--}-limits for square integrable martingales. Here is the hard part in the regularity theorem for martingales which concludes section 9. Equipped with it, the definition of two-directional projections and the proof of the above mentioned properties in section 10 is rather straightforward. In section 11 we extend the results of section 5: we give a condition under which weak submartingales decompose into weak martingales and previsible increasing processes. If, in addition, they are also submartingales, we describe the weak martingale part more precisely by a martingale and two i-submartingales which are increasing in $\bar{i}$-direction. The latter decomposition will prove to be extremely important in the following chapters.

7. Two-directional stopping; a predictable section theorem

A geometrical object which undoubtedly could be adequate for defining a "past" and a "future" is a dividing line of Π such that no two points on it are strictly comparable in the partial ordering of Π. Given such a line L, anything lying below L could be said to belong to the "past", anything above to the "future". Also, a concept of "first entrance" with respect to both time directions simultaneously is conceivable. Let S be a progressively measurable set in $\Omega \times \Pi$. Take all lines in the above sense in the future of which S lies. Intersecting all their futures gives the future of a line which could be called "line of first entrance". These ideas are basic to the notion of "stopping lines", the two-directional stopping concept we consider in this section. Like the stopping points, stopping lines are far from attaining the significance stopping times have in the theory of one-parameter processes. However, they will help us out of the dilemma that we were left with after section 6. Since previsible stopping points cannot necessarily be "predicted" by an increasing sequence of stopping points, it seems impossible to obtain the existence of Q^{--}-limits for martingales. The predictable section theorem we prove at the end of this section will help to fill this gap.

We start with stating precisely what "lines of first entrance" should be and follow Bakry [2] and Meyer [33]. See also Merzbach [28] and the references there.

Definition 1: Let $S \subset \Omega \times \Pi$. The random set $]S,(1,1)]$ defined by

$$]S,(1,1)](\omega) = \bigcup_{t \in S_\omega}]t,(1,1)], \quad \omega \in \Omega,$$

is called "open envelope" of S. If D(S) denotes the boundary of $]S,(1,1)]$ ("line of first entrance" of S), the random set $[S,(1,1)] = D(S) \cup]S,(1,1)]$ is called "closed envelope" of S. Let

$$D(S)^1 = \limsup_{n\to\infty} [D(S) + \Omega \times \{(0,1/n)\}] ,$$

$$D(S)^2 = \limsup_{n\to\infty} [D(S) + \Omega \times \{(1/n,0)\}] ,$$

$$D(S)^0 = D(S) \setminus D(S)^1 \cup D(S)^2 .$$

The random set $D(S)^i$ is called "i-part" of D(S), i=0,1,2, $D(S)^0$ also "set of exposed points" of D(S).

Remarks: 1. $D(S)^1$ ($D(S)^2$) is the union of all left open, right closed vertical (horizontal) components of D(S). As soon as we study jumps of processes it will become clear why $D(S)^i$ is called i-part of D(S).

2. Obviously, $D(S) = D(D(S)^0)$ for $S \subset \Omega \times \Pi$.

In section 3 we proved that if $S \in \mathfrak{M}^1$, its "stopping points of first entrance" are 1-optional. For lines of first entrance we have the following analogue.

Proposition 1: Let $S \in \mathfrak{M}$. Then:

1. $]S,(1,1)] \in \mathfrak{P}$, $[S,(1,1)] \in \mathfrak{G}$,
2. $D(S)$, $D(S)^o \in \mathfrak{G}$, $D(S)^1 \in \mathfrak{G} \cap \mathfrak{P}^2$, $D(S)^2 \in \mathfrak{G} \cap \mathfrak{P}^1$.

A corresponding statement holds with $\mathfrak{M}^i, \mathfrak{G}^i, \mathfrak{P}^i$ instead of $\mathfrak{M}, \mathfrak{G}, \mathfrak{P}$, $i=1,2$. Moreover, if $D(S) \in \mathfrak{P}$, then $D(S)^i \in \mathfrak{P}$ for $i=0,1,2$ and an analogous statement with $\mathfrak{P}^j$, $j=1,2$.

Proof:

1. For $t \in \mathbb{I}$ let

$$F_t = \{\omega \in \Omega: \Omega \times [0,t[\cap S \neq \emptyset\}.$$

Then $F_t = \pi_\Omega(\Omega \times [0,t[\cap S) \in \mathcal{F}_t$ by the completeness of $\mathbb{F}$. Now note that

$$]S,(1,1)] = \bigcup_{t \in \mathbb{I} \cap \mathbb{Q}^2} F_t \times]t,(1,1)],$$

and proposition (2.2) implies $]S,(1,1)] \in \mathfrak{P}$. By proposition (2.1) and theorem (2.2), $[S,(1,1)] \in \mathfrak{M}$. But the definition of $\mathfrak{G}$ shows $[S,(1,1)] \in \mathfrak{G}$. Hence also $D(S) \in \mathfrak{G}$.

2. For $i=1,2$, let Y be an $\mathbb{F}_i$-optional process on $\Omega \times [0,1]$. Then for $n \in \mathbb{N}$

$$Z^n = Y_{(.-1/n) \vee 0}$$

is $\mathbb{F}_i$-previsible. Therefore, $D(S) + \Omega \times \{(0,1/n)\}$ is $\mathfrak{P}^2$-measurable for $n \in \mathbb{N}$. This implies $D(S)^1 \in \mathfrak{G} \cap \mathfrak{P}^2$. Similarly, $D(S)^2 \in \mathfrak{G} \cap \mathfrak{P}^1$. Hence, finally, $D(S)^o \in \mathfrak{G}$. For $\mathfrak{P}$-measurability, the arguments are the same. □

Remark: If $S = S^a \in \mathfrak{P}$, then $D(S) \in \mathfrak{P}$. We do not need this statement here and refer the reader to Bakry [2], p. 68, for its difficult proof.

Definition 2: A set $L \subset \Omega \times \mathbb{I}$ is called "(weak) stopping line", if $L = D(L) \in \mathfrak{M}$ $(\mathfrak{M} \ni L \subset D(L))$.

According to definition 2, the weak stopping lines are just the progressively measurable parts of stopping lines. For stopping lines, progressive measurability just means optionality and the analogy with i-stopping points is complete.

Proposition 2: Let $L \subset \Omega \times \Pi$. Then:

1. L is a stopping line iff $L = D(L) \in \mathfrak{G}$;
2. if L is a stopping line, $L^o \in \mathfrak{G}$, $L^1 \in \mathfrak{G} \cap \mathfrak{P}^2$, $L^2 \in \mathfrak{G} \cap \mathfrak{P}^1$ are weak stopping lines; if $L \in \mathfrak{P}$, then $L^i \in \mathfrak{P}$ for $i=0,1,2$ and a similar statement holds for $\mathfrak{P}^j$, $j=1,2$.

Proof:

The assertion follows immediately from proposition 1. □

We now introduce "stochastic intervals" with respect to stopping lines.

Definition 3: Let L_1, L_2 be stopping lines. Then

$]L_1,L_2] =]L_1,(1,1)] \setminus]L_2,(1,1)]$, if $[L_2,(1,1)] \subset [L_1,(1,1)]$.

Analogously, $[L_1,L_2[$, $]L_1,L_2[$ and $[L_1,L_2]$ are defined.

Remark: Let L_1,L_2 be stopping lines such that $[L_2,(1,1)] \subset [L_1,(1,1)]$. Then by proposition 1, $]L_1,L_2] \in \mathfrak{P}$, whereas $[L_1,L_2[$, $]L_1,L_2[$, $[L_1,L_2[\in \mathfrak{G}$.

A stopping time is called predictable, if it can be strictly approximated from below by a sequence of stopping times. There is a similar notion of predictability for stopping lines.

Definition 4: A stopping line L is called "predictable", if there exists a sequence $(L_n)_{n \in \mathbb{N}}$ of stopping lines (a "predicting sequence") such that

$$L = \bigcap_{n \in \mathbb{N}}]L_n,L].$$

Remarks: 1. If L is a predictable stopping line, then $L \in \mathfrak{P}$. This is an immediate consequence of the remark after definition 3. If a stopping line $L \in \mathfrak{P}$, then L is predictable. This is proved by Bakry [2].

2. For the main subjects of the first two chapters, projection and regularity theorems, stochastic intervals of the form $]L_1,L_2]$ play the same role as the previsible neighborhoods of stopping points in section 3. To be more precise: in the proof of the existence of Q^{--}-limits for martingales, for a sequence $(L_n)_{n\in\mathbb{N}}$ of stopping lines predicting L, the previsible sets $]L_n,L]$ replace the R_n and S_n in section 3 for approximation in Q^{--}. Since previsible stopping points are not necessarily predictable by a sequence of stopping points, section 3 cannot provide an appropriate equivalent of these stochastic intervals (see Bakry [3], p. 41). There is a notion of i-predictable stopping lines as well, i=1,2, and there are analogues of the following predictable section theorem (see Bakry [2]). For our presentation, we chose to share the jobs: stopping points take care of the regularity problems in Q^{++}, Q^{+-} and Q^{-+}, stopping lines in Q^{--}. Therefore the theorem is only for the previsible case.

<u>Theorem 1</u>: Let $S\in\mathcal{P}$, $\varepsilon>0$.

Then there exists a previsible weak stopping line $L\subset S$ such that

i) $L(\omega,.)$ is closed for all $\omega\in\Omega$,

ii) $D(L)$ is predictable,

iii) $P(\pi_\Omega(S)\setminus\pi_\Omega(L))<\varepsilon$.

<u>Proof</u>:

Denote by $\mathcal{S}$ the system of finite unions of rectangles of the form $F\times[s,t]$, where $s,t\in\Pi$, $s\leq t$, $F\in\bigcup_{u<s}\mathcal{F}_u\vee\mathcal{F}_0$. Since any previsible rectangle $G\times]u,v]$ is the union of $G\times[(u+(\frac{1}{n},\frac{1}{n}))\wedge v,v]$, $n\in\mathbb{N}$, proposition (2.2) implies that $\mathcal{S}$ generates $\mathcal{R}$, hence $\mathcal{P}$. Moreover, the line of first entrance of any element of $\mathcal{S}$ is obviously predictable.

Now appeal to the classical section theorem (Dellacherie, Meyer

[17], pp. 103-105) to choose a "random point" $T \in \mathcal{F}\times\mathcal{B}(\Pi)$ such that $T \subset S$ and $\pi_\Omega(T) = \pi_\Omega(S)$. We consider the finite measure

$$m(U) = P(\pi_\Omega(U \cap T)), \quad U \in [\mathcal{F}\times\mathcal{B}(\Pi)]\vee\mathcal{N},$$

whose support is T. By a classical extension theorem, we can choose a decreasing sequence $(R_n)_{n\in\mathbb{N}}$ in $\mathcal{S}$ such that

$R = \bigcap_{n\in\mathbb{N}} R_n \subset S$ and

$$m(R) \geq m(S) - \varepsilon = P(\pi_\Omega(S)) - \varepsilon.$$

Since the corresponding sequence $(D(R_n))_{n\in\mathbb{N}}$ of predictable stopping lines is increasing, $L = D(R)\cap R$ is obviously appropriate. □

8. Martingale inequalities and applications

To consider the simplest case, the case of a bounded random variable X, will be the first step in the investigation of two-directional projections. They turn out to be M, $M^{-\cdot}$, $M^{\cdot-}$ and M^{--} in terms of M, the martingale $E(X|\mathcal{F}.)$. Therefore, the study of projections is postponed until we have fully established the regularity properties of martingales. For this sake, we need to provide some further knowledge on martingales. This will occupy the next two sections, the first one of which is devoted to the central martingale inequalities of Doob-Cairoli and Burkholder. Before we prove them, we establish on the basis of section 6 that every bounded martingale possesses a version which is Q^{++}-continuous and has Q^{+-}- and Q^{-+}-limits, a result which with the help of Doob's inequality extends to $L\log^+L$-bounded martingales. Finally, two important applications are given. We show that L^2-bounded martingales of orthogonal variation are orthogonal and prepare the decomposition theorem for M^2 by proving uniform integrability properties like the "class D^i" properties of section 5, if M is a square integrable martingale. Compare Bakry [2] and Meyer [33].

Proposition 1: Let M be a bounded martingale. Then M possesses a version which is Q^{++}-continuous and has Q^{+-}- and Q^{-+}-limits.

Proof:

Let $X = M_{(1,1)}$. We consider the iterated optional projections of X. Using conditional independence of $\mathbb{F}_1$ and $\mathbb{F}_2$ and a monotone class argument, we see that the processes ${}^{\gamma_1\gamma_2}X$ and ${}^{\gamma_2\gamma_1}X$ are progressively measurable, hence adapted. Let $t \in \Pi$, $F \in \mathcal{F}_t$ be arbitrary and set $A = 1_F \, 1_{[t,(1,1)]}$. Then A is a 1- and 2-optional integrable increasing process and an iterated application of theorem (4.1) reveals

$$E(1_F \, ({}^{\gamma_1\gamma_2}X)_t) = E(1_F \, M_t) = E(1_F \, ({}^{\gamma_2\gamma_1}X)_t).$$

Since F is arbitrary and both ${}^{\gamma_1\gamma_2}X$ and ${}^{\gamma_2\gamma_1}X$ are adapted, we see that they are versions of the martingale M. Let us turn to the regularity properties of the projections. By theorem (6.1), ${}^{\gamma_2}X$ is $Q^{\pm +}$-continuous and possesses $Q^{\pm -}$-limits. In particular, ${}^{\gamma_2}X$ is Q^{++}-continuous and possesses Q^{+-}-limits. Now by theorem (6.2), ${}^{\gamma_1\gamma_2}X$ is Q^{++}-continuous and possesses Q^{+-}-limits. In the same way, ${}^{\gamma_2\gamma_1}X$ is Q^{++}-continuous and possesses Q^{-+}-limits. By Q^{++}-continuity, ${}^{\gamma_1\gamma_2}X = {}^{\gamma_2\gamma_1}X$ is Q^{++}-continuous and possesses Q^{+-}- and Q^{-+}-limits. □

The proof of the following maximal inequalities of Doob and Cairoli is much like the proof of the projection inequalities in section 4 or section 10. In the martingale case, they are contained in the latter. For the convenience of later reference, we state them in both the one- and two-parameter setting.

Theorem 1: Let M be a Q^{++}-continuous martingale or non-negative submartingale, N a right-continuous martingale or non-negative submartingale on $\Omega\times[0,1]$. Moreover, let Φ be a co-moderate Young function, $\Psi(t) = t\log^+ t$, $t\geq 0$. Then

$$\|\sup_{r\in[0,1]} |N_r|\|_\Phi \leq \frac{p}{p-1}\|N_1\|_\Phi ,$$

$$\|\sup_{t\in\Pi} |M_t|\|_\Phi \leq (\frac{p}{p-1})^2\|M_{(1,1)}\|_\Phi .$$

In particular, for $q>1$

$$\|\sup_{r\in[0,1]} |N_r|\|_q \leq \frac{q}{q-1}\|N_1\|_q ,$$

$$\|\sup_{t\in\Pi} |M_t|\|_q \leq (\frac{q}{q-1})^2\|M_{(1,1)}\|_q .$$

Moreover, there is a constant c which does not depend on M, N such that for any $\lambda>0$

$$\|\sup_{r\in[0,1]} |N_r|\|_1 \leq c\|N_1\|_\Psi , \quad \lambda P(\sup_{r\in[0,1]} |N_r| > \lambda) \leq c\|N_1\|_1 ,$$

$$\lambda P(\sup_{t\in\Pi} |M_t| > \lambda) \leq c\|M_{(1,1)}\|_\Psi .$$

Proof:

1. Let $N^* = \sup_{r\in[0,1]} |N_r|$, $\mathbb{A}$ the set of integrable increasing processes on Π which are bounded by 1. Then by similar arguments as in the proof of theorem (4.4)

$$\|N^*\|_1 \leq \sup_{A\in\mathbb{A}} \int_{\Omega\times\Pi} |N|\, dm_A = \sup_{A\in\mathbb{A}} \int_{\Omega\times\Pi} |N|\, dm_{A^{\gamma_1}}$$

$$\leq \sup_{A\in\mathbb{A}} \int_{\Omega\times\Pi} |N_1|\, dm_{A^{\gamma_1}} = \sup_{A\in\mathbb{A}} E(|N_1| A^{\gamma_1}_{(1,1)}),$$

where for the inequality in the second line we need the submartingale property of $|N|$, and the one-parameter situation is transferred to a two-parameter setting in an obvious way. Like in the proof of theorem (4.4) we obtain

$$\|N^*\|_1 \leq c\|N_1\|_\Psi$$

with a constant c which does not depend on N. For the remaining one-parameter inequalities, the reader is referred to Dellacherie, Meyer [18], pp. 184-186.

2. Let us now derive the inequalities for two-parameter martingales from their one-parameter counterparts. To this end, let

$M^* = \sup_{t\in\Pi} |M_t|, \; R_r = \sup_{t_2\in[0,1]} |M_{(r,t_2)}|, \; r\in[0,1].$

The Q^{++}-continuity of M takes care of measurability questions. Moreover, $M^* = \sup_{r\in[0,1]} R_r$. Since for any $t_2\in[0,1]$, the process $(|M_{(t_1,t_2)}|)_{t_1\in[0,1]}$ is an $\mathbb{F}_1$-submartingale, R is a (non-negative) submartingale as well. Now apply the one-parameter inequalities twice, once on R, once on $M_{(1,.)}$. □

Our (preliminary) regularity result can be extended to $L \log^+ L$-bounded martingales.

Proposition 2: Let M be a martingale which is bounded in $L \log^+ L$. Then M possesses a version which is Q^{++}-continuous and has Q^{+-}- and Q^{-+}-limits.

Proof:

For $n\in\mathbb{N}$ consider the bounded martingale

$$M^n = E(-n\vee(M_{(1,1)}\wedge n)|\mathcal{F}.).$$

Then clearly

$$\| M^n_{(1,1)} - M_{(1,1)} \|_\Psi \underset{n\to\infty}{\to} 0,$$

where $\Psi(t) = t \log^+ t$, $t\geq 0$. Using theorem 1 and the lemma of Borel-Cantelli, we can extract a subsequence of $(M^n)_{n\in\mathbb{N}}$ which converges uniformly on Π to M. But uniform convergence preserves the asserted regularity properties. Therefore, proposition 1 allows to conclude. □

The "square function" inequalities of Burkholder-Davis which we will state next are probably the most important tool for the construction of quadratic variation, as will be seen in chapter IV. Again, we also give their one-parameter versions.

Theorem 2: Let Φ be a moderate, Ψ a moderate and co-moderate

Young function, $\Lambda(t) = t \log^+ t$, $t \geq 0$. Then there are constants $c_1, c_2, c_3 > 0$ such that for any Q^{++}-continuous martingale M (right-continuous martingale N on $\Omega \times [0,1]$), any partition $\mathbb{J}$ of Π ($[0,1]$), any $\lambda > 0$, setting $S_{\mathbb{J}}(X) = \sum_{J \in \mathbb{J}} (\Delta_J X)^2$ for $X = M, N$,

i) $c_1 \| S_{\mathbb{J}}(N)^{1/2} \|_\Phi \leq \| \sup_{r \in [0,1]} |N_r| \|_\Phi \leq c_2 \| S_{\mathbb{J}}(N)^{1/2} \|_\Phi$,

ii) $c_1 \| S_{\mathbb{J}}(M)^{1/2} \|_\Psi \leq \| \sup_{t \in \Pi} |M_t| \|_\Psi \leq c_2 \| S_{\mathbb{J}}(M)^{1/2} \|_\Psi$,

iii) $\lambda P(S_{\mathbb{J}}(M)^{1/2} > \lambda) \leq c_3 \| M_{(1,1)} \|_\Lambda$.

Moreover, there is an analogue of ii) with $M_{(1,1)}$ instead of $\sup_{t \in \Pi} |M_t|$ and a corresponding analogue of i) if Φ is in addition co-moderate.

Proof:

For the one-parameter inequalities, see Dellacherie, Meyer [18], p. 304. For a proof of iii) which uses a simple argument concerning Hilbert space valued martingales and is therefore out of our scope here, see Frangos, Imkeller [22]. We will prove the analogue of ii) with $\sup_{t \in \Pi} |M_t|$ replaced by $M_{(1,1)}$. This together with theorem 1 easily implies ii).

Fix a partition $\mathbb{J} = \mathbb{J}_1 \times \mathbb{J}_2$ of Π and a martingale M. Let $\{r_J^i : J \in \mathbb{J}_i, i=1,2\}$ be a family of independent random variables on an auxiliary probability space $(\Gamma, \mathcal{H}, Q)$ taking the values 1, -1 with probability $\frac{1}{2}$ each. For $\gamma \in \Gamma$, $\omega \in \Omega$, $i=1,2$, define

$$T_i^\gamma M(\omega) = \sum_{J_i \in \mathbb{J}_i} r_{J_i}^i(\gamma)\, \Delta_{J_i \cap [0,.] \times [0,.]} M(\omega),$$

$$T^\gamma M(\omega) = \sum_{J_1 \times J_2 \in \mathbb{J}} r_{J_1}^1(\gamma)\, r_{J_2}^2(\gamma)\, \Delta_{J \cap [0,.]} M(\omega).$$

Then $T_i^{\cdot} M$, $T^{\cdot} M$ are $\mathcal{H} \times \mathcal{F} \times \mathcal{B}(\Pi)$-measurable, $T_i^\gamma M$, $T^\gamma M$ are martingales and $T_1^\gamma T_2^\gamma M = T_2^\gamma T_1^\gamma M = T^\gamma M$, $T^\gamma T^\gamma M = M$, $i=1,2$. We apply i) and theorem 1 twice to find constants $b_1, \ldots, b_5 > 0$, not depending on M and $\mathbb{J}$ such that for $\gamma \in \Gamma$

$$(8.1)\quad \| T^{\gamma} M_{(1,1)} \|_{\Psi}$$
$$\leq b_1 \| [\sum_{J_1 \in \mathbb{J}_1} (r^1_{J_1})^2(\gamma) \, (\Delta_{J_1} T^{\gamma}_2 M_{(.,1)})^2]^{1/2} \|_{\Psi}$$
$$= b_1 \| [\sum_{J_1 \in \mathbb{J}_1} (\Delta_{J_1} T^{\gamma}_2 M_{(.,1)})^2]^{1/2} \|_{\Psi}$$
$$\leq b_2 \| \sup_{t_1 \in [0,1]} | T^{\gamma}_2 M_{(t_1,1)} | \|_{\Psi}$$
$$\leq b_3 \| T^{\gamma}_2 M_{(1,1)} \|_{\Psi}$$
$$\leq b_4 \| M_{(1,1)} \|_{\Psi} = b_4 \| T^{\gamma} T^{\gamma} M_{(1,1)} \|_{\Psi}$$
$$\leq b_5 \| T^{\gamma} M_{(1,1)} \|_{\Psi} .$$

Next, we fix $\omega \in \Omega$, apply Khinchin's lemma (see for example Burkholder [9], p. 33) to $T^{\cdot} M(\omega)$ and observe that $\{r^1_{J_1} \, r^2_{J_2} : J_1 \times J_2 \in \mathbb{J}\}$ is again a family of orthogonal random variables on $(\Gamma, \mathfrak{H}, Q)$ taking the values 1 and -1 each with probability $\frac{1}{2}$. This gives

$$(8.2)\quad b_6 \, \Psi([S_{\mathbb{J}}(M)]^{1/2}(\omega)) \leq E^Q(\Psi(|TM(.,\omega)|)) \leq b_7 \, \Psi([S_{\mathbb{J}}(M)]^{1/2}(\omega)),$$

where the positive constants b_6, b_7 do not depend on M and $\mathbb{J}$, and where E^Q is the expectation with respect to Q. It remains to integrate (8.1) with Q and (8.2) with P and to compare. □

Burkholder's inequalities will now help us to establish a relationship between orthogonality and the property of having orthogonal variation.

Proposition 3: Let M, N be square integrable martingales having orthogonal variation. Then M and N are orthogonal.

Proof:

1. Let $(\mathbb{J}_n)_{n \in \mathbb{N}}$ be a 0-sequence of partitions of Π. Then for $n \in \mathbb{N}$

$$\sum_{J \in \mathbb{J}_n} |\Delta_J M \, \Delta_J N| \leq [\sum_{J \in \mathbb{J}_n} (\Delta_J M)^2 \sum_{J \in \mathbb{J}_n} (\Delta_J N)^2]^{1/2}.$$

Choose a moderate Young function Φ such that $E(\Phi^2(|M_{(1,1)}|)) < \infty$ and $E(\Phi^2(|N_{(1,1)}|)) < \infty$ (see Burkholder, Davis, Gundy [8], p. 238). Let $\Psi = \Phi^2$. Then for $t>0$

$$\frac{t\Psi'(t)}{\Psi(t)} = 2 \, \frac{t\Phi'(t)}{\Phi(t)}$$

and therefore Ψ is moderate and co-moderate (see Dellacherie, Meyer [18], p. 178). Moreover, theorem 2 implies

$$\sup_{n\in\mathbb{N}} \| [\sum_{J\in\mathbb{J}_n} (\Delta_J X)^2]^{1/2} \|_\Psi < \infty \text{ for } X = M, N.$$

Hence by the uniform integrability criterion of de la Vallée-Poussin, $\{ \sum_{J\in\mathbb{J}_n} |\Delta_J M \, \Delta_J N| : n \in \mathbb{N}\}$ is uniformly integrable. Now M and N are of independent variation. Therefore, the theorem of Vitali implies that $(\sum_{J\in\mathbb{J}_n} |\Delta_J M \, \Delta_J N|)_{n\in\mathbb{N}}$ converges to 0 in $L^1(\Omega,\mathcal{F},P)$.

2. Fix $J =]s,t] \in \mathcal{J}$. Then

$$\begin{aligned}\Delta_J(MN) &= (\Delta_J M + \Delta_{J^1} M + \Delta_{J^2} M + M_s)(\Delta_J N + \Delta_{J^1} N + \Delta_{J^2} N + N_s) \\ &\quad - (\Delta_{J^1} M + M_s)(\Delta_{J^1} N + N_s) - (\Delta_{J^2} M + M_s)(\Delta_{J^2} N + N_s) \\ &\quad + M_s N_s \\ &= \Delta_J M \, \Delta_J N \\ &\quad + \Delta_J M \, (\Delta_{J^1} N + \Delta_{J^2} N + N_s) + \Delta_J N \, (\Delta_{J^1} M + \Delta_{J^2} M + M_s) \\ &\quad + \Delta_{J^1} M \, \Delta_{J^2} N + \Delta_{J^1} N \, \Delta_{J^2} M.\end{aligned}$$

The martingale property of M and N therefore gives

$$(8.3) \quad E(\Delta_J(MN)|\mathcal{F}_s) = E(\Delta_J M \, \Delta_J N|\mathcal{F}_s).$$

Hence for any 0-sequence $(\mathbb{J}_n)_{n\in\mathbb{N}}$ of partitions of $\mathbb{I}$ such that for any $n\in\mathbb{N}$ the partition $\mathbb{J}_n$ is finer than the one generated by J

$$E(\Delta_J(MN)|\mathcal{F}_s) = \sum_{J\supset K\in\mathbb{J}_n} E(\Delta_K M \, \Delta_K N|\mathcal{F}_s), \; n\in\mathbb{N}.$$

Apply Jensen's inequality and part 1 to obtain the weak martingale property of MN. This completes the proof. □

The following application of Burkholder's inequalities is the key for the decomposition theorem for the square of square integrable martingales.

<u>Proposition 4</u>: Let M be a square integrable martingale. Then:

1. for any 0-sequence $(\mathbb{J}_n)_{n \in \mathbb{N}}$ of partitions of $[0,1]$ the family $\left(\sum_{J \in \mathbb{J}_n} E((\Delta_J M_{(\cdot,1)})^2 \mid \mathcal{F}^1_{s^J}) : n \in \mathbb{N}\right)$ is uniformly integrable,

and an analogous statement holds for the second parameter,

2. for any 0-sequence $(\mathbb{K}_n)_{n \in \mathbb{N}}$ of partitions of Π the family $\left(\sum_{K \in \mathbb{K}_n} E((\Delta_K M)^2 \mid \mathcal{F}_{s^K}) : n \in \mathbb{N}\right)$ is uniformly integrable, and

analogous statements hold for conditioning with $\mathbb{F}_1$ and $\mathbb{F}_2$ instead of $\mathbb{F}$.

Proof:

We prove the more complicated second assertion. Fix a 0-sequence $(\mathbb{K}_n)_{n \in \mathbb{N}}$ of partitions of Π. For $n \in \mathbb{N}$ consider the filtration $\mathbb{F}^n$ defined by

$$\mathcal{F}^n_t = \mathcal{F}_{t^K} \quad \text{if } t \in [s^K, t^K[, \quad K \in \mathbb{K}_n,$$

and the increasing process

$$A^n_t = \sum_{[0,t] \supset K \in \mathbb{K}_n} (\Delta_K M)^2, \qquad t \in \Pi.$$

Then, obviously,

$$(A^n)^{\pi_1 \pi_2}_{(1,1)} = \sum_{K \in \mathbb{K}_n} E((\Delta_K M)^2 \mid \mathcal{F}_{s^K}), \qquad n \in \mathbb{N}.$$

Like in the proof of proposition 3 choose a moderate and comoderate Young function Ψ which satisfies $E(\Psi(|M_{(1,1)}|)) < \infty$. Now apply theorem (4.3) twice and theorem 2 to find a constant c such that

$$\sup_{n \in \mathbb{N}} \Big\| \sum_{K \in \mathbb{K}_n} ((\Delta_K M)^2 \mid \mathcal{F}_{s^K}) \Big\|_\Psi \leq p^2 \sup_{n \in \mathbb{N}} \Big\| \sum_{K \in \mathbb{K}_n} (\Delta_K M)^2 \Big\|_\Psi$$
$$\leq c \| M_{(1,1)} \|_\Psi .$$

An appeal to the uniform integrability criterion of de la Vallée - Poussin finishes the proof. □

9. A stochastic integral and the regularity of martingales

This section is mainly devoted to the completion of the regularity results for $L \log^+ L$-bounded martingales: the existence of Q^{--}-limits. At the same time, this will provide the basis for the two-directional projection and decomposition theorems.

Our presentation is oriented along the following ideas. To gain control of the behavior of $\underline{M}^{--}$ and $\overline{M}^{--}$ we have to know what "stopping a martingale M at a stopping line" L means. Since L can be a rather complicated geometrical object, this concept does not necessarily make sense for general processes. For square integrable M, the stochastic integral of M over the stochastic interval [0,L] can be considered as M, stopped at L. Therefore, prior to the desired regularity theorem there must be a study of the stochastic integral of M. To define it, we will use the standard isometry argument resting upon the martingale property of M and the existence of the Doléans measure of M^2. More precisely, before we can define the integral of M, we have to verify that m_{M^2} exists using a uniform integrability criterion like in proposition (5.1). But this essentially amounts to proving the "existence" part of a decomposition theorem first. Strangely enough, this forces us to prove "existence" in order to prepare the ground for the proof of "uniqueness" in the decomposition of M^2. This presentation is partly due to Bakry [1] and Meyer [33].

The existence of the Doléans measure of M^2 is treated with the methods of section 5.

<u>Proposition 1</u>: Let M be a square integrable martingale. Then:
1. M^2 is a submartingale and a weak submartingale,

2. the Doléans measure of M^2 exists.

Proof:

1. Let $J =]s,t] \in \mathcal{J}$ be given. By some elementary algebra involving the conditional independence property

(9.1) $E(M_t^2|\mathcal{F}_s) = E([(\Delta_J M)^2 + (\Delta_{J^1} M)^2 + (\Delta_{J^2} M)^2]|\mathcal{F}_s) + M_s^2$

and by (8.3), applied to M=N,

(9.2) $E((\Delta_J M)^2|\mathcal{F}_s) = E(\Delta_J M^2|\mathcal{F}_s)$.

(9.1) implies the submartingale property, (9.2) the weak submartingale property of M^2.

2. Let $X = M^2$, $(\mathbb{II}_n)_{n\in\mathbb{N}}$ a 0-sequence of partitions of $\mathbb{I}$. Then by (9.2) for any $n\in\mathbb{N}$

$$\sum_{J\in\mathbb{II}_n} E(\Delta_J X|\mathcal{F}_{s^J}) = \sum_{J\in\mathbb{II}_n} E((\Delta_J M)^2|\mathcal{F}_{s^J}).$$

Now proposition (8.4) applies and shows that $\{\sum_{J\in\mathbb{II}_n} E(\Delta_J X|\mathcal{F}_{s^J}): n\in\mathbb{N}\}$ is uniformly integrable. Moreover, X is $\hat{Q}^{++}$-continuous in $L^1(\Omega,\mathcal{F},P)$. The arguments of part 1 of the proof of proposition (5.1) now work and the existence of the Doléans measure follows. □

We are now in a position to construct a stochastic integral of M.

Definition 1: The linear space $\mathcal{E}$, generated by the indicator functions of previsible rectangles, is called space of "previsible elementary processes".

Proposition 2: Let M be a square integrable martingale. Then the linear mapping

$$\cdot M : \quad \mathcal{E} \to L^2(\Omega,\mathcal{F},P),$$

$$Y_o = \sum_{1\le i\le n} a_i 1_{F_i\times J^i} \to \sum_{1\le i\le n} a_i 1_{F_i} \Delta_{J^i} M = Y_o\cdot M,$$

has a linear extension to $L^2(\Omega\times\mathbb{I},\mathcal{P},m_{M^2})$ which is an isometry.

Proof:

Note first that any previsible elementary process has a representation of the form $Y_o = \sum_{J\in\mathbb{J}} \alpha_J 1_J$, where $\mathbb{J}$ is a partition of Π, $\alpha_J \in \mathfrak{M}(\mathcal{F}_{s_J}, \mathcal{B}(\mathbb{R}))$ a step function for $J\in\mathbb{J}$. The martingale property of M and (9.2) give

$$(9.4) \quad \| Y_o\cdot M\|_2^2 = \| \sum_{J\in\mathbb{J}} \alpha_J^2 (\Delta_J M)^2 \|_1 = \| \sum_{J\in\mathbb{J}} \alpha_J^2 \Delta_J M^2 \|_1 = \int_{\Omega\times\Pi} Y_o^2 \, dm_{M^2}.$$

Now a standard completion argument finishes the proof. □

Definition 2: Let M be a square integrable martingale. The linear extension according to proposition 2 is called "stochastic integral" of M and denoted by $\cdot M$.

In a similar way, integral processes can be defined.

Proposition 3: Let M be a Q^{++}-continuous square integrable martingale. Then the linear mapping

$$\cdot M. : \mathcal{E} \to L^{2,\infty}(\Omega\times\Pi, [\mathcal{F}\times\mathcal{B}(\Pi)]\vee\mathcal{N}, \mathcal{B}(\mathbb{R})),$$

$$Y_o \to (Y_o 1_{\Omega\times[0,.]})\cdot M = Y_o\cdot M. ,$$

has a linear extension to $L^2(\Omega\times\Pi, \mathcal{P}, m_{M^2})$ which satisfies

i) $\| \sup_{t\in\Pi} |Y\cdot M_t| \|_2 \le 4[\int_{\Omega\times\Pi} Y^2 \, dm_{M^2}]^{1/2}$,

ii) $Y\cdot M.$ is a Q^{++}-continuous martingale which possesses Q^{+-}- and Q^{-+}-limits if M does and is continuous if M is,

iii) $Y\cdot M_t = (Y 1_{\Omega\times[0,t]})\cdot M$,

for any $Y\in L^2(\Omega\times\Pi, \mathcal{P}, m_{M^2})$, $t\in\Pi$.

Proof:

Note first that for $Y_o\in\mathcal{E}$ the process $Y_o\cdot M.$ is a martingale which inherits all the regularity properties of M. Therefore, up to a standard argument, we have to show that inequality i) holds for all $Y_o\in\mathcal{E}$. This, however, follows from theorem (8.1) and (9.4). □

Definition 3: Let M be a Q^{++}-continuous square integrable martingale. The integral process according to proposition 3 is called "stochastic integral process" of M and denoted by $\cdot M.$. For $S \in \mathcal{P}$ the martingale $1_S \cdot M.$ is also written $M(S)$.

The following "localization" property of integral processes will be valuable.

Proposition 4: Let M be a Q^{++}-continuous square integrable martingale, L a stopping line and Y, Z bounded Q^{--}-continuous processes such that $\{Y > 0\} \subset [L,(1,1)]$, $\{Z > 0\} \subset [0,L]$. Then

$$1_{[0,L[}\ Y \cdot M. = 0, \quad 1_{]L,(1,1)]}\ \Delta_{].,(1,1)]}\ Z \cdot M. = 0.$$

Proof:

Fix $t \in \mathrm{I\!I}$ and a 0-sequence $(\mathcal{J}_n)_{n \in \mathbb{N}}$ of partitions of $\mathrm{I\!I}$. For $n \in \mathbb{N}$ set

$$Y^n = \sum_{J \in \mathcal{J}_n} Y_{s^J}\, 1_J .$$

The sequence $(Y^n)_{n \in \mathbb{N}}$ of previsible bounded processes converges in $L^2(\Omega \times \mathrm{I\!I}, \mathcal{P}, m_{M^2})$ to Y and for any $n \in \mathbb{N}$ we have

$$1_{[0,L[} \sum_{J \in \mathcal{J}_n} Y_{s^J}\, \Delta_{J \cap [0,.]} M = 0.$$

Therefore, the first conclusion can be drawn with the help of propositions 2 and 3. The second one is easier. Note that for $t \in \mathrm{I\!I}$

$$\{t \in]L,(1,1)]\} \times]t,(1,1)] \in \mathcal{R}.$$

Therefore

$$1_{]L,(1,1)]}(t)\ \Delta_{]t,(1,1)]} Z \cdot M. = (1_{]L,(1,1)]}(t)\ 1_{]t,(1,1)]}\ Z) \cdot M$$
$$= 0$$

by iii) of proposition 3, for $t \in \mathrm{I\!I}$. The same proposition allows to conclude. □

The preceding three propositions tell us how to stop a martingale at a stopping line and the last one provides enough information on the stopped martingale to enable us to attack the problem of the existence of Q^{--}-limits.

Proposition 5: Let M be a square integrable martingale. Then M possesses a regular version.

Proof:

Following proposition (8.2) we may assume that M is Q^{++}-continuous and possesses Q^{+-}- and Q^{-+}-limits. We have to show that M has Q^{--}-limits. Assume the set

$$\{\underline{M}^{--} < \overline{M}^{--}\},$$

which is previsible according to theorem (2.1), is non-evanescent. Choose $a,b \in \mathbb{R}$, $a<b$, such that

$$S = \{\underline{M}^{--} < a,\ b < \overline{M}^{--}\} \in \mathcal{P}$$

is still non-evanescent. Theorem (7.1) produces a previsible weak stopping line $K \subset S$ whose line of first entrance $L = D(K)$ is predictable and such that $P(\pi_\Omega(K)) > 0$. Let $(L_n)_{n \in \mathbb{N}}$ be a predicting sequence for L. Since L is previsible, all the sets $[0,L[$, $[0,L_n]$, $]L_n,L[$ are previsible, $n \in \mathbb{N}$. Since $]L_n,L[\downarrow \emptyset$, proposition 3,i) allows us to assume

$$(9.5) \quad \sup_{t \in \Pi} |M(]L_n,L[)_t| \underset{n\to\infty}{\to} 0.$$

Now observe that for any $t \in \Pi$

$$1_{\Pi} + 1_{[0,t]} = 1_{[0,(t_1,1)]} + 1_{[0,(1,t_2)]} + 1_{]t,(1,1)]}.$$

Hence for any previsible set T, any $t \in \Pi$

$$M(T)_{(1,1)} + M(T)_t = M(T)_{(t_1,1)} + M(T)_{(1,t_2)} + \Delta_{]t,(1,1)]}M(T).$$

We apply this formula to $T = [0,L[$, observing that in consequence of proposition 4 the process $1_{[0,L[}\, M([0,L[)$ equals $1_{[0,L[}\, M$, whereas the process $1_{]L_n,(1,1)]}\, \Delta_{].,(1,1)]}\, M([0,L_n])$ is evanescent, $n \in \mathbb{N}$. This gives

$$1_{]L_n,L[}[M([0,L[)_{(1,1)} + M]$$
$$= 1_{]L_n,L[}\,[M([0,L[)_{(.,1)} + M([0,L[)_{(1,.)} + \Delta_{].,(1,1)]}M([0,L[)],$$

$n \in \mathbb{N}$.

We subtract this equation for s, $t \in \mathbb{I}$ and estimate to obtain

(9.6) $1_{]L_n,L[}(s) \; 1_{]L_n,L[}(t) \; |M_s - M_t|$

$$\leq 1_{]L_n,L[}(s) \; 1_{]L_n,L[}(t) \; [|M([0,L[)_{(s_1,1)} - M([0,L[)_{(t_1,1)}|$$

$$+ |M([0,L[)_{(1,s_2)} - M([0,L[)_{(1,t_2)}|]$$

$$+ \; 8 \sup_{t \in \mathbb{I}} \; |M(]L_n,L[)_t|, \quad n \in \mathbb{N}.$$

Now pick $\omega \in \pi_\Omega(K)$ satisfying (9.5), $u \in K_\omega$ and two sequences $(s^m)_{m \in \mathbb{N}}$, $(t^m)_{m \in \mathbb{N}}$ converging to u in $Q^{--}(u)$ and such that $M_{s^m}(\omega) \leq a$,

$M_{t^m}(\omega) \geq b$ for $m \in \mathbb{N}$. Moreover, by (9.5) choose n large enough to

ensure $\sup_{t \in \mathbb{I}} \; |M(]L_n,L[)_t(\omega)| < (b-a)/16$. Then for almost all $m \in \mathbb{N}$

(9.6) implies

$$(b-a)/2 \leq |M([0,L[)_{(s_1^m,1)} - M([0,L[)_{(t_1^m,1)}|(\omega)$$

$$+ \; |M([0,L[)_{(1,s_2^m)} - M([0,L[)_{(1,t_2^m)}|(\omega).$$

Since $P(\pi_\Omega(K)) > 0$, this violates the almost sure existence of left hand limits for the one-parameter martingales $M([0,L[)_{(1,.)}$ and $M([0,L[)_{(.,1)}$ (see Dellacherie, Meyer [18], p. 76). This completes the proof.   □

Remark: It is natural to ask whether the methods of proof of proposition 5 cannot be generalized to the other quadrants to give Q^{++}-, Q^{+-}-, and Q^{-+}-regularity, replacing the methods of section 6. The main problem one has to face hereby consists in finding "sections" of the sets $S^j = \{\underline{M}^j \leq a, b \leq \overline{M}^j\}$, $j \in \{+,-\}^2$, which possess the appropriate measurability properties. The line of first entrance of S^{++} is a stopping line, since $S^{++} \in \mathbb{m}$, and therefore the method applies indeed to the quadrant Q^{++} (see Bakry [1] and Meyer [33]). Now consider for example S^{+-}. We know $S^{+-} \in \mathbb{m}^1 \cap \mathbb{p}^2$. If we take a section of S^{+-} by applying a section theorem, it might not be 1-optional. If, however, we take the line of first entrance, it might not be 2-previsible. Since, on the other hand, the methods of section 6 do not generalize to yield Q^{--}-limits, the regularity proof for martingales as a whole leaves some questions open.

Theorem 1: Let M be an L $\log^+$L-integrable martingale. Then M possesses a regular version.

Proof:

Use proposition 5 in the arguments of the proof of proposition (8.2). □

The space of square integrable martingales can be endowed with a Hilbert space norm which is uniform in the time variable.

Theorem 2: The function $M \to \| M_{(1,1)} \|_2$ defines a Hilbert space norm on the linear space of square integrable regular martingales which is equivalent to $\| \cdot \|_{2,\infty}$.

Proof:

By theorem (8.1) for any regular square integrable martingale

$$\| \sup_{t \in \Pi} |M_t| \|_2 \leq 4 \| M_{(1,1)} \|_2 \leq 4 \| \sup_{t \in \Pi} |M_t| \|_2 .$$

This implies the assertion. □

Definition 4: The Hilbert space of square integrable regular martingales is denoted by $\mathcal{M}^2$.

10. Two-directional projection theorems

Let X be a bounded random variable. We know that its 1-optional projection is the 1-martingale, which is generated by taking conditional expectations with respect to $\mathbb{F}_1$. The first step in our program of defining two-directional projections consists in studying the 2-optional projection of ${}^{\gamma_1}X$. In the proof of proposition (8.1) we already recognized it as the martingale M which equals X at time (1,1). With the support of the regularity theorem of section 9 we will be able to compute the other iterated projections in terms of M: $M^{-\cdot}$ is the 1-previsible, 2-optional, $M^{\cdot -}$ the 1-optional, 2-previsible and M^{--} the 1-and 2-previsible projection. This statement, however, implicitly says that iterated projections commute. We extend these commutation results to general bounded measurable processes and define "optional" and "previsible" projections as the product of two one-directional ones in an arbitrary order. As a by-product, we can show that the optional (previsible) sets are just the intersections of the 1-and 2-optional (previsible) sets. It is then easy to obtain the corresponding results for dual projections, and norm inequalities extending the ones proved in section 4 readily follow. Our presentation is like Bakry's [2]. See also Doléans, Meyer [19] and Merzbach, Zakai [29].

<u>Theorem 1</u>: Let $X \in \mathfrak{m}(\mathcal{F}, \mathcal{B}(\mathbb{R}))$ be bounded. Then:

1. ${}^{\gamma_1\gamma_2}X = {}^{\gamma_2\gamma_1}X \in \mathfrak{m}(\mathcal{G}, \mathcal{B}(\mathbb{R}))$ is a regular version of the martingale $E(X|\mathcal{F}.)$,

2. ${}^{\pi_1\gamma_2}X = {}^{\gamma_2\pi_1}X = Y^{-\cdot} \in \mathfrak{m}(\mathcal{P}^1 \cap \mathcal{G}^2, \mathcal{B}(\mathbb{R}))$,

   ${}^{\gamma_1\pi_2}X = {}^{\pi_2\gamma_1}X = Y^{\cdot -} \in \mathfrak{m}(\mathcal{G}^1 \cap \mathcal{P}^2, \mathcal{B}(\mathbb{R}))$,

   ${}^{\pi_1\pi_2}X = {}^{\pi_2\pi_1}X = Y^{--} \in \mathfrak{m}(\mathcal{P}, \mathcal{B}(\mathbb{R}))$,

where $Y = {}^{\gamma_1\gamma_2}X = {}^{\gamma_2\gamma_1}X$.

<u>Proof</u>:

1. The proof of proposition (8.1) shows that $Y = {}^{\gamma_1\gamma_2}X = {}^{\gamma_2\gamma_1}X$ is a Q^{++}-continuous version of the martingale $E(X|\mathcal{F}.)$ which possesses Q^{+-}- and Q^{-+}-limits. Therefore, theorem (9.1) applies and assertion 1 follows.

2. Since Y^{--} is Q^{--}-continuous, the last measurability result follows by making an appeal to theorem (2.1) . The remaining

two are easy consequences of the equations to be proved. Let us consider the first pair. Assume for example that the first of the two sets

$S^+ = \{{}^{\pi_1\gamma_2}X < Y^{-\cdot}\}, \quad S^- = \{Y^{-\cdot} < {}^{\pi_1\gamma_2}X\}$

which are 1-previsible by theorem (2.1), is non-evanescent. Use theorem (3.1) to choose a previsible 1-stopping point T such that, setting $F = \{T \subset S^+\}$,

$T = (1,1)$ on $\bar{F}$, $P(F) > 0$.

Apply theorem (4.1) to the integrable increasing process $A = 1_F\, 1_{[T,(1,1)]} \in \mathfrak{M}(\mathcal{P}^1, \mathcal{B}(\mathbb{R}))$ to obtain

(10.1) $E(1_F\, {}^{\pi_1\gamma_2}X_T) = E(1_F\, {}^{\gamma_2}X_T)$.

On the other hand, for $t_2 \in [0,1]$, $G \in \mathcal{F}^1_{T_1-}$, the 1-martingale property of Y implies

$E(1_G\, (Y_{(1,t_2)} - Y^{-\cdot}_{(T_1,t_2)}) \mid \mathcal{F}^1_{T_1-}) = 0$

(see Dellacherie, Meyer [18], p. 99). By approximating T_2 from above using $\mathcal{F}^1_{T_1-}$ -measurable step functions and regularity, this generalizes to

$E(Y_{(1,T_2)} - Y^{-\cdot}_T \mid \mathcal{F}^1_{T_1-}) = 0.$

Therefore

(10.2) $E(1_F\, Y^{-\cdot}_T) = E(1_F\, Y_{(1,T_2)}) = E(1_F\, {}^{\gamma_2}X_T)$.

But (10.1) and (10.2) clearly contradict $P(F) > 0$. An analogous argument for S^- implies ${}^{\pi_1\gamma_2}X = Y^{-\cdot}$. To prove ${}^{\gamma_2\pi_1}X = Y^{-\cdot}$, consider the sets

$\{{}^{\gamma_2\pi_1}X < Y^{-\cdot}\}$ and $\{Y^{-\cdot} < {}^{\gamma_2\pi_1}X\}$

which are 2-optional by regularity of Y and argue similarly, choosing at first a 2-stopping point instead of a previsible 1-stopping point. Also, the last pair of equations poses no new problem. To prove that ${}^{\pi_1\pi_2}X = Y^{--}$, we can take over the above reasoning with only one minor change: T_2 has to be approximated from below by $\mathcal{F}^1_{T_1-}$ -measurable step functions. This completes the proof. □

The definition of general two-directional projections is now almost a pure matter of monotone class arguments.

Proposition 1: Let $X \in \mathfrak{M}([\mathcal{F}\times\mathcal{B}(\mathbb{I})]\vee\mathcal{N}, \mathcal{B}(\mathbb{R}))$ be bounded, A an integrable increasing process. Then:

1. ${}^{\gamma_1\gamma_2}X = {}^{\gamma_2\gamma_1}X \in \mathfrak{M}(\mathcal{G}, \mathcal{B}(\mathbb{R}))$, $\quad A^{\gamma_1\gamma_2} = A^{\gamma_2\gamma_1} \in \mathfrak{M}(\mathcal{G}, \mathcal{B}(\mathbb{R}))$,
2. ${}^{\pi_1\gamma_2}X = {}^{\gamma_2\pi_1}X \in \mathfrak{M}(\mathcal{P}^1\cap\mathcal{G}^2, \mathcal{B}(\mathbb{R}))$, $\quad A^{\pi_1\gamma_2} = A^{\gamma_2\pi_1} \in \mathfrak{M}(\mathcal{P}^1\cap\mathcal{G}^2, \mathcal{B}(\mathbb{R}))$,
3. ${}^{\gamma_1\pi_2}X = {}^{\pi_2\gamma_1}X \in \mathfrak{M}(\mathcal{G}^1\cap\mathcal{P}^2, \mathcal{B}(\mathbb{R}))$, $\quad A^{\gamma_1\pi_2} = A^{\pi_2\gamma_1} \in \mathfrak{M}(\mathcal{G}^1\cap\mathcal{P}^2, \mathcal{B}(\mathbb{R}))$,
4. ${}^{\pi_1\pi_2}X = {}^{\pi_2\pi_1}X \in \mathfrak{M}(\mathcal{P}, \mathcal{B}(\mathbb{R}))$, $\quad A^{\pi_1\pi_2} = A^{\pi_2\pi_1} \in \mathfrak{M}(\mathcal{P}, \mathcal{B}(\mathbb{R}))$.

In particular, $\mathcal{G} = \mathcal{G}^1\cap\mathcal{G}^2$ and $\mathcal{P} = \mathcal{P}^1\cap\mathcal{P}^2$.

Proof:

Once the results for the projections are proved, the dual projections can be treated by applying theorems (4.1) and (4.2) twice and proposition (2.3).

Let $X = Y\, 1_J$, where J is an interval in $\mathbb{I}$, $Y \in \mathfrak{M}(\mathcal{F}, \mathcal{B}(\mathbb{R}))$ bounded. Then for $(\alpha_1,\alpha_2) \in \{\gamma_1,\gamma_2,\pi_1,\pi_2\}^2$

$$ {}^{\alpha_1\alpha_2}X = {}^{\alpha_1\alpha_2}Y\, 1_J . $$

Therefore, 1.-4. are true for processes of this simple kind. Now monotone class arguments have to be applied.

Finally, assume $S \in \mathcal{G}^1\cap\mathcal{G}^2$. Then

$$ 1_S = {}^{\gamma_1\gamma_2}1_S \in \mathfrak{M}(\mathcal{G}, \mathcal{B}(\mathbb{R})) $$

in consequence of part 1. Hence $S \in \mathcal{G}$. This implies $\mathcal{G}^1\cap\mathcal{G}^2 \subset \mathcal{G}$. The opposite inclusion is trivial. The arguments for the previsible sets are the same. □

Remark: Given the results of theorem 1 and proposition 1, it is easy to show that $\mathcal{G}^1\cap\mathcal{P}^2$ is generated by the adapted processes which are Q^{+-}-continuous and possess limits in the remaining three quadrants. Analogous results hold for any pair in $\{\mathbb{G}^1,\mathbb{P}^1\}\times\{\mathbb{G}^2,\mathbb{P}^2\}$. Remember that $\mathbb{G} = \mathbb{G}^1\cap\mathbb{G}^2$ was defined to be generated by the adapted regular processes.

Theorem 2: Let $X \in \mathfrak{M}([\mathcal{F}\times\mathcal{B}(\Pi)]\vee\mathcal{N}, \mathcal{B}(\mathbb{R}))$ be bounded, $\mathbb{A} \in \{\mathbb{G}^1,\mathbb{P}^1\}$, $\mathbb{B} \in \{\mathbb{G}^2,\mathbb{P}^2\}$. Then there exists a unique process $Y \in \mathfrak{M}(\mathbb{A}\cap\mathbb{B}, \mathcal{B}(\mathbb{R}))$ such that for any integrable increasing process $A \in \mathfrak{M}(\mathbb{A}\cap\mathbb{B}, \mathcal{B}(\mathbb{R}))$

$$\int_{\Omega\times\Pi} X\, dm_A = \int_{\Omega\times\Pi} Y\, dm_A .$$

Moreover, Y is given by the respective iterated projections of proposition 1.

Proof:

"Existence" is settled by proposition 1 and theorem (4.1). To prove uniqueness, denote by α,β the projection symbols associated with $\mathbb{A}$, $\mathbb{B}$. Assume Y, Z are $\mathbb{A}\cap\mathbb{B}$-measurable and satisfy

$$\int_{\Omega\times\Pi} Y\, dm_A = \int_{\Omega\times\Pi} X\, dm_A = \int_{\Omega\times\Pi} Z\, dm_A$$

for any integrable increasing process $A \in \mathfrak{M}(\mathbb{A}\cap\mathbb{B}, \mathcal{B}(\mathbb{R}))$.

Consider

$$U = \{Y < Z\} \in \mathbb{A}\cap\mathbb{B}, \quad V = \{Z < Y\} \in \mathbb{A}\cap\mathbb{B}$$

and assume for example that U is non-evanescent. Choose a stopping point $T \in \mathbb{A}$ such that $T = (1,1)$ on $\{T\not\subset U\}$ and $P(T\subset U) > 0$. This is guaranteed by theorem (3.1). Let $A = 1_{\{T\subset U\}}\, 1_{[T,(1,1)]}$. Then A is an integrable increasing process in $\mathfrak{M}(\mathbb{A}, \mathcal{B}(\mathbb{R}))$. By proposition 1, A^β is in $\mathfrak{M}(\mathbb{A}\cap\mathbb{B}, \mathcal{B}(\mathbb{R}))$. Hence

$$E(1_{\{T\subset U\}}(Z_T-Y_T)) = \int_{\Omega\times\Pi} (Z-Y)\, dm_{A^\beta} = 0$$

which contradicts $P(T\subset U) > 0$. An analogous argument for V yields the desired equality of Y and Z. □

Definition 1: Let $X \in \mathfrak{M}([\mathcal{F}\times\mathcal{B}(\Pi)]\vee\mathcal{N}, \mathcal{B}(\mathbb{R}))$ be a bounded process. The process Y which exists according to theorem 2 is called in the respective cases "optional projection", "1-optional, 2-previsible projection", "1-previsible, 2-optional projection", "previsible projection" of X. The optional (previsible) projection of X is also denoted by ${}^{\gamma}X$ (${}^{\pi}X$).

For dual projection we have the following result.

Theorem 3: Let A be an integrable increasing process, $\mathfrak{A}\in\{\mathfrak{G}^1,\mathfrak{P}^1\}$, $\mathfrak{B}\in\{\mathfrak{G}^2,\mathfrak{P}^2\}$. Then there exists a unique integrable increasing process $B \in \mathfrak{m}(\mathfrak{A}\cap\mathfrak{B}, \mathfrak{B}(\mathbb{R}))$ such that for any bounded process $X \in \mathfrak{m}(\mathfrak{A}\cap\mathfrak{B}, \mathfrak{B}(\mathbb{R}))$

$$\int_{\Omega\times\Pi} X\, dm_A = \int_{\Omega\times\Pi} X\, dm_B.$$

Moreover, B is given by the respective iterated dual projections of proposition 1.

Proof:

An appeal to proposition 1 and theorem (4.2) is enough for "existence". Uniqueness follows from theorem 2, applied twice, and proposition (2.3). □

Definition 2: Let A be an integrable increasing process. The integrable increasing process B according to theorem 3 is called in the respective cases "dual optional projection", " dual 1-optional, 2-previsible projection", "dual 1-previsible, 2-optional projection", "dual previsible projection" of A. The dual optional (previsible) projection of A is also denoted by A^{γ} (A^{π}).

The norm inequalities of section 4 for (dual) projections now generalize in an obvious way.

Theorem 4: Let A be an integrable increasing process, Φ a moderate function, B one of the dual projections of A according to definition 2. Then

$$\| B_{(1,1)} \|_{\Phi} \leq p^2 \| A_{(1,1)} \|_{\Phi}.$$

In particular, for $q\geq 1$

$$\| B_{(1,1)} \|_{q} \leq q^2 \| A_{(1,1)} \|_{q}.$$

Proof:

Apply theorem (4.3) twice. □

Theorem 5: Let $X \in \mathfrak{M}([\mathcal{F}\times\mathcal{B}(\Pi)]\vee N, \mathcal{B}(\mathbb{R}))$ be bounded, Φ a co-moderate Young function and $\Psi(t) = t(\log^+ t)^2$, $t \geq 0$, Y one of the projections of X according to theorem 2. Then

$$\| \sup_{t\in\Pi} |Y_t| \|_\Phi \leq 4\left(\frac{p}{p-1}\right)^2 \| \sup_{t\in\Pi} |X_t| \|_\Phi .$$

In particular, for $q>1$

$$\| \sup_{t\in\Pi} |Y_t| \|_q \leq 4\left(\frac{q}{q-1}\right)^2 \| \sup_{t\in\Pi} |X_t| \|_q .$$

Moreover, there exists a constant c which does not depend on X such that

$$\| \sup_{t\in\Pi} |Y_t| \|_1 \leq c \| \sup_{t\in\Pi} |X_t| \|_\Psi .$$

Proof:

Apply theorem (4.4) twice to obtain the first part. For the second one, an argument as in the proof of the corresponding inequality in theorem (4.4) works. The derivative of the conjugate function of Ψ is bounded by a constant multiple of

$$t \to \exp((1+t)^{1/2}), \quad t \geq 0.$$

But theorem 4 implies that there is a constant c such that

$$\sup_{A\in\mathbf{A}} E(\exp([A^{\alpha_1\alpha_2}]^{1/2}/c)) < \infty,$$

where $\mathbf{A}$ is the set of all integrable increasing processes bounded by 1 and $\alpha_1, \alpha_2 \in \{\gamma_1,\gamma_2,\pi_1,\pi_2\}$. This gives the desired result. □

11. Two-directional decomposition theorems

Our treatment of two-directional decompositions already started in proposition 1 of section 9, where the existence of the Doléans measure of M^2 for a square integrable martingale M has been established. The criterion we used there was a two-directional version of the "class D^i" property of section 5 and will play just that role in the general theory, too. The property, called "of class D", will firstly prove to be equivalent to the existence of the Doléans measure of a weak submartingale X which therefore possesses a unique decomposition by a weak martingale and a previsible integrable increasing process. By far more important to our main decomposition theorems of martingales in chapters III and IV will, however, be an extension of the decomposition derived in theorem (5.2). If X is both a weak submartingale and a submartingale in the order sense which is of class D and D^i, i=1,2, the 2-submartingale A which shows up in theorem (5.2) turns out to be of class D^2 and therefore decomposes further. We obtain a representation of X by a martingale, a sum of i-submartingales which are of bounded variation in $\bar{i}$-direction and an increasing process. By a simple iteration argument we again get norm inequalities relating X and its "dual projections". Since the square of a square integrable martingale is of class D and D^i, i=1,2, by proposition (8.4), we can finally complete its decomposition started in section 9. For our methods see Brennan [5], Cairoli [11] and Merzbach [28] as well as the references there, for the results also Wong, Zakai [46].

Definition 1: Let X be a process such that X_t is integrable for any $t \in \Pi$. X is said to be "of class D", if for any 0-sequence $(\mathbb{J}_n)_{n \in \mathbb{N}}$ of partitions of Π the family

$$\{ \sum_{J \in \mathbb{J}_n} E(\Delta_J X | \mathcal{F}_{s_J}) \; : \; n \in \mathbb{N}\}$$

is uniformly integrable.

Proposition 1: Let X be a weak submartingale which is $\hat{Q}^{++}$-continuous in $L^1(\Omega,\mathcal{F},P)$. Then m_X exists iff X is of class D.

Proof:

In the proof of proposition (5.1), replace the appeal to theorem (4.4) by theorem (10.5) which gives an estimation of ${}^{\gamma}1_F$ for $F \in \mathcal{F}$. Otherwise, the arguments are strictly analogous. □

Proposition 1 yields the following decomposition of weak submartingales by weak martingales and previsible increasing processes.

Theorem 1: Let X be a weak submartingale of class D which is $\hat{Q}^{++}$-continuous in $L^1(\Omega,\mathcal{F},P)$. Then there exists a unique integrable increasing process $A \in \mathfrak{M}(\mathcal{P}, \mathcal{B}(\mathbb{R}))$ such that X-A is a weak martingale, or such that $m_X|\mathcal{P} = m_A|\mathcal{P}$. Moreover, for any 0-sequence $(\mathbb{J}_n)_{n\in\mathbb{N}}$ of partitions of Π, any $t\in\Pi$

$$A_t = \lim_{n\to\infty} \sum_{J\in\mathbb{J}_n} E(\Delta_{J\cap[0,t]}X|\mathcal{F}_{s^J}) \quad \text{weakly in } L^1(\Omega,\mathcal{F},P).$$

Proof:

For the existence and uniqueness of A the arguments of the proof of theorem (5.1) work, with theorem (10.3) replacing theorem (4.2). Let $(\mathbb{J}_n)_{n\in\mathbb{N}}$ be a 0-sequence of partitions of Π and take $t = (1,1)$. Pick $F\in\mathcal{F}$ and denote by M a regular version of the martingale $E(1_F|\mathcal{F}.)$. Then for any $n\in\mathbb{N}$

$$E(1_F \sum_{J\in\mathbb{J}_n} E(\Delta_J X|\mathcal{F}_{s^J})) = E(\sum_{J\in\mathbb{J}_n} M_{s^J}\,\Delta_J X)$$

$$= E(\sum_{J\in\mathbb{J}_n} M_{s^J}\,\Delta_J A).$$

Therefore by theorems (10.1) and (10.2)

$$\lim_{n\to\infty} E(1_F \sum_{J\in\mathbb{J}_n} E(\Delta_J X|\mathcal{F}_{s^J})) = \int_{\Omega\times\Pi} M^{--}\, dm_A = E(1_F\, A_{(1,1)}).$$

This allows to conclude. □

Definition 2: Let X be a weak submartingale of class D which is $\hat{Q}^{++}$-continuous in $L^1(\Omega,\mathcal{F},P)$. The integrable increasing process A according to theorem 1 is called "dual previsible projection" of X.

In case X is not only a weak submartingale, but also a submartingale in the order sense and of class D^i, i=1,2, the weak martingale appearing in the representation of theorem 1 can be given a more precise description.

Theorem 2: Let X be a weak submartingale and a submartingale of class D, D^1 and D^2 which is $\hat{Q}^{++}$-continuous in $L^1(\Omega,\mathcal{F},P)$. Let A^i resp. A be its dual i-previsible resp. previsible projections according to theorem (5.2) resp.theorem 1, i=1,2. Then:

1. A^i is of class $D^{\bar{i}}$ and A is the dual $\bar{i}$-previsible projection of A^i,
2. $M^i = A^{\bar{i}} - A$ is an i-martingale, $M = X - A^1 - A^2 + A$ a martingale, i=1,2,
3. $X = M + M^1 + M^2 + A$.

Proof:

Let $(\mathbb{J}_n)_{n\in\mathbb{N}}$ be a 0-sequence of partitions of [0,1]. Then by theorem (5.2) for any $t\in\mathbb{I}$

$$A^1_t = \lim_{n\to\infty} \sum_{J\in\mathbb{J}_n} E(\Delta_{J\cap[0,t_1]\times[0,t_2]}X|\mathcal{F}^1_{s^J}) \text{ weakly in } L^1(\Omega,\mathcal{F},P).$$

This immediately implies that A^1 is $\hat{Q}^{++}$-continuous in $L^1(\Omega,\mathcal{F},P)$. To prove that A^1 is of class D^2, let $(\mathbb{K}_m)_{m\in\mathbb{N}}$ be another 0-sequence of partitions of [0,1]. Then for any $m\in\mathbb{N}$

$$\sum_{K\in\mathbb{K}_m} E(\Delta_K A^1_{(1,.)}|\mathcal{F}^2_{s^K}) = \lim_{n\to\infty} \sum_{J\times K\in\mathbb{J}_n\times\mathbb{K}_m} E(\Delta_{J\times K}X|\mathcal{F}_{(s^J,s^K)})$$

weakly in $L^1(\Omega,\mathcal{F},P)$. Since X is of class D,

$$\{ \sum_{J\times K\in\mathbb{J}_n\times\mathbb{K}_m} E(\Delta_{J\times K}X|\mathcal{F}_{(s^J,s^K)}) : n,m\in\mathbb{N} \}$$

is uniformly integrable. Therefore, A^1 is of class D^2. Let B be the dual 2-previsible projection of A^1 according to theorem (5.1). Since for any $r\in[0,1]$, $A^1_{(.,r)}$ is the dual previsible projection of $X_{(.,r)}$, a monotone class argument starting with the indicator functions of 2-previsible rectangles shows that

$$\int_{\Omega\times\mathbb{I}} Y\, dm_{A^1} = \int_{\Omega\times\mathbb{I}} {}^{\pi_1}Y\, dm_{A^1}$$

for bounded $Y \in \mathfrak{M}(\mathcal{P}^2, \mathcal{B}(\mathbb{R}))$. This entails the same equation for B instead of A^1. Now proposition (4.1) implies that B, besides being 2-previsible, is also 1-previsible, hence previsible. To sum up, B is a previsible integrable increasing process and

$$X - B = X - A^1 + A^1 - B$$

is the sum of a 1-and a 2-martingale, consequently a weak martingale. Theorem 1 therefore gives B = A. The arguments for A^2 being parallel, part 1 follows. All that remains to be done to prove parts 2 and 3 is to show that M is a martingale. But since

$$M = (X - A^1) - (A^2 - A) = (X - A^2) - (A^1 - A),$$

M is both a 1- and a 2-martingale, hence a martingale by the conditional independence property of $\mathbb{F}$. □

The most important examples of processes which possess a decomposition according to theorem 2 are the integrable increasing processes and the squares of square integrable martingales.

<u>Corollary 1</u>: Let A be an integrable increasing process. Then

1. $M^i = A^{\pi_{\bar{i}}} - A^{\pi}$ is an i-martingale, $M = A - A^{\pi_1} - A^{\pi_2} + A^{\pi}$ a martingale,
2. $A = M + M^1 + M^2 + A^{\pi}$.

<u>Corollary 2</u>: Let M be a square integrable martingale. Then M^2 possesses a decomposition according to theorem 2. Moreover, if $\langle M\rangle^i$ resp. $\langle M\rangle$ denotes the dual i-previsible resp. previsible projection of M^2, for any 0-sequence $(\mathbb{J}_n)_{n\in\mathbb{N}}$ of partitions of [0,1] resp. $\mathbb{I}$, any $t \in \mathbb{I}$

$$\langle M\rangle^i_t = \lim_{n\to\infty} \sum_{J\in\mathbb{J}_n} E((\Delta_{J\cap[0,t_i]} M(\cdot,t_{\bar{i}}))^2 | \mathcal{F}^i_{s_J}), \quad i=1,2,$$

$$\langle M\rangle_t = \lim_{n\to\infty} \sum_{J\in\mathbb{J}_n} E((\Delta_{J\cap[0,t]} M)^2 | \mathcal{F}_{s_J}), \text{ weakly in } L^1(\Omega,\mathcal{F},P).$$

Proof:

The statement that M^2 is a submartingale and a weak submartingale is contained in proposition (9.1). Proposition (8.4) shows that M^2 is of class D, D^1 and D^2. For this and the two representation results remember (9.2). Theorem 2 takes care of the first part of the assertion, theorem (5.2) and theorem 1 of the representations of the dual projections of M^2. □

Definition 3: Let M be a square integrable martingale. The dual i-previsible resp. previsible projection of M^2 according to corollary 2 is called "i-previsible process" resp. "previsible process" associated with M and denoted by $\langle M\rangle^i$ resp. $\langle M\rangle$, i=1,2.

We return to the general theory to state Garsia-Neveu type inequalities for dual previsible projections.

Theorem 3: Let $X \in \mathfrak{M}([\mathcal{F}\times\mathcal{B}(\mathbb{I})]\vee\mathcal{N}, \mathcal{B}(\mathbb{R}))$ be a weak submartingale and a submartingale of class D, D^1 and D^2 which is $\hat{Q}^{++}$-continuous in $L^1(\Omega,\mathcal{F},P)$, A its dual previsible projection, Φ a moderate and co-moderate Young function. Then

$$\| A_{(1,1)} \|_\Phi \le 4p^2 \frac{p}{p-1} \| \sup_{t_i\in[0,1]} |X_{(t_i,1)}| \|_\Phi \quad , \ i=1,2.$$

In particular, for q>1

$$\| A_{(1,1)} \|_q \le 4q^2 \frac{q}{q-1} \| \sup_{t_i\in[0,1]} |X_{(t_i,1)}| \|_q \quad , \ i=1,2.$$

If X is nonnegative, the factor 4 on the right hand side may be replaced by 1.

Proof:

Let for example A^1 be the dual 1-previsible projection of X according to theorem (5.2). Since X is $\hat{Q}^{++}$-continuous in $L^1(\Omega,\mathcal{F},P)$, we may assume that $A^1_{(1,.)}$ is right-continuous (see Dellacherie, Meyer [18], p. 76). Now by the proof of theorem 2 and theorem (5.3)

$$\| A_{(1,1)} \|_\Phi$$

$$\le 2p \, \| \sup_{t_2 \in [0,1]} A^1_{(1,t_2)} \|_\Phi$$

$$\le 4p^2 \frac{p}{p-1} \, \| \sup_{t_1 \in [0,1]} |X_{(t_1,1)}| \, \|_\Phi \, .$$

An analogous argument with A^2 instead of A^1 gives the desired inequality for $i=2$. □

Remark: The problem of whether weak submartingales of class D satisfy an inequality like in theorem 3 has not been considered.

We finally specialize the Garsia-Neveu inequalities of section 5 and theorem 3 for the processes associated with a square integrable martingale.

Corollary: Let $M \in \mathcal{M}^2$, $i=1,2$, Φ a moderate and co-moderate Young function. Then

$$\| \sup_{t_{\bar{i}} \in [0,1]} \langle M\rangle^i_{(1,t_{\bar{i}})} \|_\Phi \le p\left(\frac{p}{p-1}\right)^2 \, \| M^2_{(1,1)} \|_\Phi \, ,$$

$$\| \langle M\rangle_{(1,1)} \|_\Phi \le p^2 \left(\frac{p}{p-1}\right)^2 \, \| M^2_{(1,1)} \|_\Phi \, .$$

In particular, for $q>1$

$$\| \sup_{t_{\bar{i}} \in [0,1]} \langle M\rangle^i_{(1,t_{\bar{i}})} \|_q \le q\left(\frac{q}{q-1}\right)^2 \, \| M^2_{(1,1)} \|_q \, ,$$

$$\| \langle M\rangle_{(1,1)} \|_q \le q^2 \left(\frac{q}{q-1}\right)^2 \, \| M^2_{(1,1)} \|_q \, .$$

Proof:

Combine the inequalities of theorem (5.3), theorem 3 and theorem (8.1). □

III. Jumps of martingales and their compensation

Equipped with the projection and decomposition theorems of chapters I and II, we can now turn to establish the foundation of the main theorems which describe the structure of square integrable martingales and their quadratic variation. We already know that square integrable martingales possess regular versions. Therefore we can reasonably talk about their i-jumps, i=0,1,2. The subject of the structure theorem is to gradually extract the jumps of such a martingale M, decomposing it into "i-jump parts", i=0,1,2, and a continuous part, which are also martingales. This is executed in three steps, beginning with the 0-jumps on which we concentrate in the following considerations. In a first attempt to extract them one could start by defining the "0-jump part" of M at t to be the sum of all jumps of M up to t. But unless M is previsible, this process is not a martingale and therefore needs to be "compensated". This means that we have to find a process C (a "compensator") such that, if ΔM denotes the 0-jump part of M, the process ΔM-C is a martingale and such that C has no 0-jumps. Otherwise, the extraction of ΔM-C from M would eventually not remove all 0-jumps. To find a compensator, we apply the decomposition theorems of the first two chapters to ΔM. But ΔM need not be of bounded variation or of a submartingale type. We therefore have to define jump parts on smaller discontinuity sets and divide the process of extraction into countably many steps. The sets L which prove to be appropriate for our purposes are called "simple sets" and have the following structure: L is optional and its Π-sections are finite sets of points (for 0-jumps) or finite unions of vertical line segments (for 1-jumps) or finite unions of horizontal line segments (for 2-jumps) and is accordingly called "i-simple", i=0,1,2. For a 0-simple set L, the results of chapters I and II can be applied to the process M(L) of 0-jumps of M on L. They yield a process C such that M(L)-C is indeed a martingale. But, unless L is specified further, C may have 0-jumps. To realize, how L has to be specified, let us look at M(L) more closely. One can decompose M(L) into two increasing processes, the positive jumps $\overline{M}(L)$ and the negative jumps $\underline{M}(L)$. As is plausible from corollary 2 of theorem (11.2), the favourite candidate for C has to be the process

$$[\overline{M}(L)^{\pi_1}+\overline{M}(L)^{\pi_2}-\overline{M}(L)^{\pi}]-[\underline{M}(L)^{\pi_1}+\underline{M}(L)^{\pi_2}-\underline{M}(L)^{\pi}].$$

Which properties of L could make this process be free of 0-jumps? If L is previsible in direction i, M(L) behaves like a martingale in this direction and therefore $\overline{M}(L)^{\pi_i}=\underline{M}(L)^{\pi_i}$, $\overline{M}(L)^{\pi}=\underline{M}(L)^{\pi}$. If L is "inaccessible" in direction $\overline{i}$, i.e. for every 0-simple $\overline{i}$-previsible set K, the intersection of K and L is a.s. trivial, the processes $\overline{M}(L)^{\pi_i}$, $\underline{M}(L)^{\pi_i}$, being $\overline{i}$-previsible, cannot see the jump set L and therefore are continuous in $\overline{i}$-direction. In particular, they have no 0-jumps. Our first aim consequently has to be to decompose the 0-jump set of M into 0-simple sets which are combinations of "i-previsible", "i-inaccessible" and "$\overline{i}$-previsible", "$\overline{i}$-inaccessible". The simple fact that L can be studied by means of the "associated increasing process" $\Gamma(L)$ which, at time t, just counts the points in $[0,t]\cap L$, is very helpful: the investigation of the jumps of the dual previsible projections of $\Gamma(L)$ enables us to decompose L into unique parts of the various accessibility degrees.

In this chapter we run the program just sketched up to the compensation of martingale jumps on simple sets. In section 12 we introduce i-simple sets, i=0,1,2, and study their relationship with stopping lines on one hand and - more importantly - with increasing processes on the other hand. The latter is exploited in section 13, where we prove that 0-simple sets can be decomposed by "totally inaccessible", "1-previsible, 2-inaccessible", "2-previsible, 1-inaccessible" and previsible ones, 1- and 2-simple sets by "inaccessible" and "previsible" ones. In section 14, jump processes of regular martingales on simple sets are defined. Their elementary properties are considered and a decomposition of the jump sets of a regular martingale by countably many simple sets of the different kinds discussed in section 13 is given. Finally, in section 15, we show how the jump processes of martingales on simple sets of these types can be compensated in a way to make the successive extraction of jump parts from martingales possible.

12. Simple sets and increasing processes

The development of the concept studied in this section started with the question "Which are the simplest geometric objects in $\Omega \times \Pi$ still appropriate for describing the sets of discontinuities of regular processes like increasing processes or martingales?" In the theory of one-parameter processes the answer is clear: the graphs of stopping times. The set of jumps of a right-continuous process possessing left limits can be embedded in a countable union of them. Taking into account the results of section 1, it is not hard to guess what could be candidates in the two-parameter case: optional sets whose Π-sections consist of a finite number of points (for the 0-jumps) resp. vertical line segments (for the 1-jumps) resp. horizontal line segments (for the 2-jumps). Let us call them for a moment "simple sets of type I". Of course, they are not necessarily contained in graphs of stopping lines. One might therefore suspect that simple sets of type I are still too complicated and, using stopping lines, one could design a simpler type. Call a stopping line simple, if the set of its exposed points has finite Π-sections. Now a natural suggestion could be: take the i-parts (for the i-jumps) of simple stopping lines instead. Let us call them "simple sets of type II". But there is an essential difference between one- and two-parameter theory which explains why type II is indeed not simpler than type I. The fact which makes the difference between the counterparts of type I and II in the one-parameter theory is easy to grasp. The $[0,1]$-sections of the graph of a stopping time contain exactly one point, whereas the Π-sections of the counterparts of simple sets of type I are only finite. This difference disappears completely in the theory of two-parameter processes and it often appears unnatural to require simple sets to lie on graphs of stopping lines. We therefore investigate type I sets and point out only at the very beginning how they are related to graphs of stopping lines. A simple set can be covered by a countable family of them such that its Π-sections are contained in finitely many of them.

Our main interest, however, is the study of the close relationship between increasing processes and simple sets. We show that the set of i-jumps of an increasing process A can be covered by a countable union of simple sets of the respective kind which inherit the measurability properties of A, $i=0,1,2$. Conversely, given a simple set S, there is an increasing process which has the same measurability properties as S and whose discontinuity set is S. For the decomposition theorem of increasing processes compare Mazziotto, Merzbach, Szpirglas [27].

Definition 1: Let

$S^o = \{L \in \mathfrak{G}: \ \omega \to |L_\omega|$ is integrable$\}$,

$S^1 = \{L \in \mathfrak{G}: \ L_\omega$ is a finite set of vertical open line segments whose upper boundary is on $\partial\Pi$ for $\omega\in\Omega$, $\omega \to |\pi_{\Omega,1}(L)_\omega|$ is integrable$\}$,

$S^2 = \{L \in \mathfrak{G}: \ L_\omega$ is a finite set of horizontal open line segments whose right boundary is on $\partial\Pi$ for $\omega\in\Omega$, $\omega \to |\pi_{\Omega,2}(L)_\omega|$ is integrable$\}$.

Then S^i is called the class of "i-simple" sets. A set $L\in S^i$ is called "p-integrable", if $\omega \to |L_\omega|$ resp. $|\pi_{\Omega,i}(L)_\omega|$ is p-integrable, $p\geq 1$, $i=0,1,2$. For $i=1,2$, denote by $s(S^i)$ the semiring generated by S^i.

Sometimes, sets in $s(S^i)$ are called i-simple as well.

Remark: S^o is a ring.

The relationship between simple sets and stopping lines is clarified by the following proposition.

Proposition 1: Let $i=0,1,2$, $L \in S^i$. Then there exists a sequence $(L_n)_{n\in\mathbb{N}}$ of pairwise disjoint weak stopping lines such that

i) $L_n = D(L_n^i) \cap L_n \subset L$ for $n \in \mathbb{N}$,

ii) $L = \bigcup_{n\in\mathbb{N}} L_n$.

Moreover, $L_n \in \mathfrak{P}^{\bar{i}}$ for $i=1,2$, and $L_n \in \mathfrak{P}$ resp. $\mathfrak{P}^j$ if $L \in \mathfrak{P}$ resp. $\mathfrak{P}^j$, $j=1,2$, $n \in \mathbb{N}$.

Proof:

Let us consider the case $i=1$. For $i=0$, the arguments are similar but easier, for $i=2$ parallel. Let $L_1 = D(L)^1$. By proposition (7.1),

$$L \supset L_1 = L\setminus]L,(1,1)] \in \mathfrak{P}^2.$$

Define $K_1 = L \setminus L_1$. Again, $K_1 \in S^1$.

Now suppose that 2-previsible pairwise disjoint weak stopping lines $L_1,\dots,L_n \in \mathcal{S}^1$ have been chosen and that $K_n = L \setminus \bigcup_{1\le i\le n} L_i$ is 1-simple. Let $L_{n+1} = D(K_n)^1$. Then again by proposition (7.1),

$L \supset L_{n+1} = K_n \setminus]K_n,(1,1)] \in \mathcal{P}^2$,

and $K_{n+1} = K_n \setminus L_{n+1} \in \mathcal{S}^1$. This recursive definition obviously yields a sequence $(L_n)_{n\in\mathbb{N}}$ of pairwise disjoint weak stopping lines satisfying i) and ii). Moreover, since open envelopes of (j-) progressively measurable sets are (j-) previsible by proposition (7.1), it also gives the remaining previsibility assertions, j=1,2. □

<u>Corollary</u>: Let i=1,2, $L \in \mathcal{S}^i$. Then $L \in \mathcal{P}^{\bar{i}}$.

Given a 0-simple set K, one can consider the set of all open vertical or horizontal line segments in Π connecting the points of K to the upper or right boundary of Π. The resulting random sets are 1-simple resp. 2-simple. Dually, given a 1- or 2-simple set L, the set of boundary points of L is a 0-simple set.

<u>Definition 2</u>:

1. For $K \in \mathcal{S}^o$ the random set $\sigma_1 K$ of vertical open line segments in Π connecting the points of K to the upper boundary of Π is called "<u>1-shadow of K</u>". Analogously, the "<u>2-shadow of K</u>" is defined and denoted by $\sigma_2 K$.
2. For i=1,2, $L \in \mathcal{S}^i$, the random set ∂L of boundary points of the line segments constituting L is called "<u>boundary of L</u>".

For shadows and boundaries the following measurability statements hold.

<u>Proposition 2</u>: Let $K \in \mathcal{S}^o$, $L \in \mathcal{S}^1$. Then:

1. $\sigma_1 K \in \mathcal{S}^1$, and $K \in \mathcal{P}^1$ implies $\sigma_1 K \in \mathcal{P}^1$,
2. $\partial L \in \mathcal{S}^o$, and $L \in \mathcal{P}^1$ implies $\partial L \in \mathcal{P}^1$.

A corresponding statement holds with $\mathcal{S}^1$, $\mathcal{P}^1$ replaced by $\mathcal{S}^2$, $\mathcal{P}^2$.

Proof:

1. Let $(K_n)_{n\in\mathbb{N}}$ be a decomposition of K by a sequence of weak stopping lines according to proposition 1. Let $L_n = D(K_n)^1$, $n\in\mathbb{N}$. Then $L_n \in \mathfrak{G}$ by proposition (7.1) for any $n\in\mathbb{N}$ and therefore $\sigma_1 K = \bigcup_{n\in\mathbb{N}} L_n \in \mathfrak{G}$. Moreover, still by proposition (7.1), $L_n \in \mathfrak{P}^1$ if $K_n \in \mathfrak{P}^1$ for $n\in\mathbb{N}$. Hence an appeal to proposition 1 completes the proof of part 1.

2. Let $(L_n)_{n\in\mathbb{N}}$ be a decomposition of L according to proposition 1. This time, let $K_n = D(L_n)^0$, $n\in\mathbb{N}$. Again by proposition (7.1), $K_n \in \mathfrak{G}$ for any $n\in\mathbb{N}$ and therefore

$$\partial L = \bigcup_{n\in\mathbb{N}} K_n \setminus L \in \mathfrak{G}.$$

For the previsibility assertion, we need a new argument. Assume $L\in\mathfrak{P}^1$. Then $L_n \in \mathfrak{P}^1$ for all $n\in\mathbb{N}$ by proposition 1. Fix $n\in\mathbb{N}$. Then

$$K_n = \limsup_{m\to\infty} [L_n - \Omega\times\{(0,\tfrac{1}{m})\}] \setminus L \in \mathfrak{P}^1.$$

Hence $\partial L \in \mathfrak{P}^1$ and the proof is complete. □

Let us next point out where simple sets originate. Roughly speaking, i-simple sets are the random sets of i-jumps of adapted integrable increasing processes, i=0,1,2. To make this statement more precise, we first investigate the jumps of a given integrable increasing process.

Proposition 3: Let $p\geq 1$, $A\in\mathfrak{M}(\mathfrak{G}, \mathfrak{B}(\mathbb{R}))$ a p-integrable increasing process. Then there exists a sequence $(L_n)_{n\in\mathbb{N}}$ of pairwise disjoint p-integrable 0-simple sets such that A has no 0-jumps outside $\bigcup_{n\in\mathbb{N}} L_n$.

Moreover, if $A\in\mathfrak{M}(\mathfrak{P}, \mathfrak{B}(\mathbb{R}))$, then also $L_n\in\mathfrak{P}$ for $n\in\mathbb{N}$. An analogous statement holds for $\mathfrak{P}^j$ instead of $\mathfrak{P}$, j=1,2.

Proof:

For $\varepsilon\geq 0$ let

$$L^0(\varepsilon) = \{(\omega,t)\in\Omega\times\mathbb{I}: \Delta_t A(\omega) > \varepsilon\}.$$

By regularity and theorem (2.1), $\Delta.A \in \mathfrak{M}(\mathfrak{G}, \mathcal{B}(\mathbb{R}))$ and even $\Delta.A \in \mathfrak{M}(\mathfrak{P}, \mathcal{B}(\mathbb{R}))$ if A is previsible. Hence $L^o(\varepsilon) \in \mathfrak{G}$ and even $L^o(\varepsilon) \in \mathfrak{P}$ if A is previsible, for all $\varepsilon \geq 0$. Moreover, for $\varepsilon > 0$, $\omega \in \Omega$

$$|L^o(\varepsilon)_\omega| \leq \frac{1}{\varepsilon} A_{(1,1)}(\omega)$$

which implies that $\omega \to |L^o(\varepsilon)_\omega|$ is p-integrable for $\varepsilon > 0$. Therefore,

$$L_n = L^o(\tfrac{1}{n}) \setminus L^o(\tfrac{1}{n-1}) \quad , \ n \in \mathbb{N},$$

is appropriate. □

If A has no 0-jumps, the 1-and 2-jumps can be controlled in a similar way.

Proposition 4: Let $p \geq 1$, $A \in \mathfrak{M}(\mathfrak{G}, \mathcal{B}(\mathbb{R}))$ a p-integrable increasing process without 0-jumps. Then there exist sequences $(K_n)_{n \in \mathbb{N}}$ of sets in $s(S^1)$ and $(L_n)_{n \in \mathbb{N}}$ of sets in $s(S^2)$ which are pairwise disjoint and p-integrable such that A has no 1-jumps outside $\bigcup_{n \in \mathbb{N}} K_n$ and no 2-jumps outside $\bigcup_{n \in \mathbb{N}} L_n$.
Moreover, if $A \in \mathfrak{M}(\mathfrak{P}, \mathcal{B}(\mathbb{R}))$, then also K_n, $L_n \in \mathfrak{P}$ for $n \in \mathbb{N}$. An analogous statement holds for $\mathfrak{P}^j$ instead of $\mathfrak{P}$, $j=1,2$.

Proof:

For $\varepsilon \geq 0$, $i=1,2$, let

$$L^i(\varepsilon) = \{(\omega,t) \in \Omega \times \mathrm{II}: \ \Delta^i_{t_i} A_{(.,t_{\bar{i}})}(\omega) > \varepsilon\}.$$

Like in the proof of proposition 3, $L^i(\varepsilon) \in \mathfrak{G}$ and $L^i(\varepsilon) \in \mathfrak{P}$ if A is previsible. Since A is increasing and has no 0-jumps,

$$t_{\bar{i}} \to \Delta^i_{t_i} A_{(.,t_{\bar{i}})}$$

is increasing and continuous, $t_i \in [0,1]$. Therefore, $L^i(\varepsilon) \in S^i$ and again

$$K_n = L^1(\tfrac{1}{n}) \setminus L^1(\tfrac{1}{n-1}), \qquad L_n = L^2(\tfrac{1}{n}) \setminus L^2(\tfrac{1}{n-1}), \quad n \in \mathbb{N},$$

is appropriate. □

By means of the preceding two propositions, the following decomposition theorem for increasing processes can be proved.

Theorem 1: Let $p \geq 1$, $A \in \mathfrak{M}(\mathfrak{G}, \mathfrak{B}(\mathbb{R}))$ a p-integrable increasing process. Then there exist unique p-integrable increasing processes A^o, A^1, A^2, $A^c \in \mathfrak{M}(\mathfrak{G}, \mathfrak{B}(\mathbb{R}))$ such that

i) for any sequence $(L_n^i)_{n \in \mathbb{N}}$ of sets in $s(S^i)$ according to propositions 3 and 4

$$A^o = \sum_{n \in \mathbb{N}} A(L_n^o), \quad A^i = \sum_{n \in \mathbb{N}} (A-A^o)(L_n^i), \ i=1,2,$$

ii) A^i has no 0-jumps and no $\bar{i}$-jumps for $i=1,2$, A^c is continuous,

iii) the random measures induced by A^o, A^1, A^2, A^c are pairwise orthogonal,

iv) $A = A^o + A^1 + A^2 + A^c$.

Moreover, $A^i \in \mathfrak{M}(\mathfrak{P}^{\bar{i}}, \mathfrak{B}(\mathbb{R}))$ for $i=1,2$, $A^c \in \mathfrak{M}(\mathfrak{P}, \mathfrak{B}(\mathbb{R}))$.
If $A \in \mathfrak{M}(\mathfrak{P}, \mathfrak{B}(\mathbb{R}))$, then also A^o, A^1, $A^2 \in \mathfrak{M}(\mathfrak{P}, \mathfrak{B}(\mathbb{R}))$. An analogous statement holds for $\mathfrak{P}^j$ instead of $\mathfrak{P}$, $j=1,2$.

Proof:

Combine propositions 3 and 4. For the previsibility results, consult proposition (2.4), the corollary of proposition 1 and the definitions of $\mathfrak{P}$, $\mathfrak{P}^1$ and $\mathfrak{P}^2$. □

Remark: Of course, we could have defined $A^o = A(L^o(0))$, $A^i = (A-A^o)(L^i(0))$, $i=1,2$, where $L^j(0)$ is like in the proofs of propositions 3 and 4, in i) of theorem 1. However, for the sake of analogy with the corresponding decomposition result for square integrable martingales in chapter IV we preferred the presentation given.

Definition 3: Let A be an integrable increasing process. The increasing process A^i according to theorem 1 is called "i-part" or "i-jump part" of A, $i=0,1,2$, A^c is called "continuous part" of A.

Now assume an i-simple set L is given. We will show that there exists an increasing process which is identical with its i-part and which has exactly the jumps prescribed by L.

Definition 4:

1. For $L \in S^o$ let

$$\Gamma(L)_t(\omega) = |L_\omega \cap [0,t]|, \quad \omega\in\Omega, \ t\in\Pi.$$

2. For $i=1,2$, $L \in S^i$ let

$$\Gamma(L)_t(\omega) = \int_{[0,t]} \lambda(]s_{\bar{i}},t_{\bar{i}}])\, d\Gamma(\partial L)_s(\omega), \ \omega\in\Omega, \ t\in\Pi .$$

By additivity, $\Gamma(L)$ is defined for $L \in s(S^i)$.

The process $\Gamma(L)$ is called "increasing process associated with L".

Remark: For $i=1,2$, $L \in S^i$, the process $\Gamma(L)$ at (ω,t) is just the sum of the lengths of the intersections of $[0,t]$ with the line segments of which L_ω consists, $(\omega,t) \in \Omega \times \Pi$.

Theorem 2: Let $i=0,1,2$, $p \geq 1$, $L \in s(S^i)$ a p-integrable set. Then

i) $\Gamma(L)$ is an optional p-integrable increasing process,

ii) $\Gamma(L) = 1_L \cdot \Gamma(L). = \Gamma(L)^i$.

Moreover, $\Gamma(L) \in \mathfrak{M}(\mathcal{P}, \mathcal{B}(\mathbb{R}))$ if L is previsible. An analogous statement holds for $\mathcal{P}^j$ instead of $\mathcal{P}$, $j=1,2$.

Proof:

1. Let us first argue for $L \in S^o$. For $\omega\in\Omega$, $\Gamma(L)_{(1,1)}(\omega) = |L_\omega|$. Hence $\Gamma(L)$ is p-integrable. By definition, $\Gamma(L)$ is increasing and regular and ii) is fulfilled. Since L is progressively measurable, $\Gamma(L)$ is adapted and therefore optional.

Now assume $L \in \mathcal{P}$. Take $B = \Gamma(L)^\pi$. We know from theorem (10.3) that $m_B(\bar{L}) = 0$. Consider the previsible sets

$$S^+ = \{(\omega,t)\in \Omega\times\Pi \ : \ \Delta_t B(\omega) > 1\},$$

$$S^- = \{(\omega,t) \in \Omega\times\Pi \ : \ \Delta_t B(\omega) < 1\}.$$

If S^+ or S^- is non-evanescent, we obtain

$$m_{\Gamma(L)}(S^+) < m_B(S^+) \quad \text{or} \quad m_{\Gamma(L)}(S^-) > m_B(S^-),$$

both in contradiction with theorem (10.3). Hence $B = \Gamma(L)$.

2. Let us consider $L \in S^1$ now. It is clear again that $\Gamma(L)$ is increasing, regular and that ii) is fulfilled. For $\omega \in \Omega$, $\Gamma(L)_{(1,1)}(\omega) \leq |\partial L_\omega|$ and p-integrability follows. Part 1 implies that $\Gamma(\partial L)$ is optional and, together with proposition 2, that $\Gamma(\partial L) \in \mathfrak{M}(\mathfrak{P}^1, \mathfrak{B}(\mathbb{R}))$ if L is 1-previsible. Moreover, by the continuity of $t_2 \to \lambda([0,t_2])$,

$\Gamma(L)$ is $\mathfrak{G}^2$- and therefore $\mathfrak{P}^2$-measurable.

By proposition (2.4) we obtain

$\Gamma(L)$ is $\mathfrak{G}^1$-measurable and $\mathfrak{P}^1$-measurable if $L \in \mathfrak{P}^1$. □

13. The decomposition of simple sets

Let T be the graph of a stopping time. One can subtract from T the graphs of countable many previsible stopping times such that the remainder has only trivial intersections with previsible stopping times. The latter part is called "inaccessible part". This decomposition can be seen to be unique (see Dellacherie, Meyer [17], pp. 214-217).

We will now solve the counterpart of this problem for the different kinds of simple sets. The main idea of our approach is to consider the increasing process $\Gamma(L)$ associated with a given simple set L and make use of our knowledge of dual previsible projections of $\Gamma(L)$. To describe the most important features of the procedure we are following in a specific example, assume $L \in S^o$. First subtract from L the union of a decomposition of the support of $[\Gamma(L)^\pi]^o$ according to theorem (12.1). Call this part "accessible". Next subtract from the remainder a corresponding part which belongs to $[\Gamma(L)^{\pi_1}]^o$ and call it "1-accessible, 2-inaccessible", and mutatis mutandis a "2-accessible, 1-inaccessible" part. The remainder after these two steps deserves the name "totally inaccessible". The uniqueness of the dual previsible projections of $\Gamma(L)$ also implies the uniqueness of this decomposition and a characterization of the inaccessible parts by a property saying that they have only trivial intersections with the respective previsible simple sets. In a similar way, simple sets of the other two kinds are treated.

The decompositions we obtain are not only of interest in their own right. They are the key to the structure theorems for square integrable martingales via "compensation of jumps" on simple sets (see sections 15 and 19). A first attempt of a definition of accessibility and inaccessibility has been made in Mazziotto, Merzbach, Szpirglas [27].

We define inaccessibility in geometrical terms.

Definition 1:

1. Let $L \in S^o$. Then:

   i) L is called "totally inaccessible", if for any 1- or 2-previsible $K \in S^o$

   $$P(\pi_\Omega(K \cap L)) = 0,$$

   ii) L is called "1-previsible, 2-inaccessible", if $L \in \mathcal{P}^1$ and for any previsible $K \in S^o$

   $$P(\pi_\Omega(K \cap L)) = 0.$$

   Similarly, the notion "2-previsible, 1-inaccessible" is defined.

2. Let $L \in s(S^1)$. Then L is called "inaccessible", if for any previsible $K \in S^1$

   $$P(\pi_\Omega(K \cap L)) = 0.$$

Analogously, inaccessibility is defined for $L \in s(S^2)$.

Our first aim is to give characterizations of the different

notions of inaccessibility of simple sets L in terms of the increasing processes associated with L. They turn out as continuity properties of the dual previsible projections of $\Gamma(L)$. The following auxiliary result will be helpful. It tells us which smoothness properties of increasing processes are inherited by their dual projections.

Proposition 1: Let $A, B \in \mathfrak{M}(\mathfrak{G}, \mathcal{B}(\mathbb{R}))$ be integrable increasing processes, i=1,2. Then:

1. m_{A^o} and $m_{(A^{\pi_i})^o}$ agree on the set of i-previsible sets in S^o,

   m_{A^o} and $m_{(A^{\pi})^o}$ agree on the set of previsible sets in S^o,

2. m_{A^1} and $m_{(A^{\pi_i})^1}$ agree on the set of i-previsible sets in $s(S^1)$, if $A^o = 0$,

   m_{A^1} and $m_{(A^{\pi})^1}$ agree on the set of previsible sets in $s(S^1)$, if $A^o = 0$,

   and an analogous statement for A^2 instead of A^1,

3. $m_{A^o} \ll m_{B^o}$ implies $m_{(A^{\pi_i})^o} \ll m_{(B^{\pi_i})^o}$ and $m_{(A^{\pi})^o} \ll m_{(B^{\pi})^o}$,

4. $m_{A^1} \ll m_{B^1}$ implies $m_{(A^{\pi_i})^1} \ll m_{(B^{\pi_i})^1}$ and $m_{(A^{\pi})^1} \ll m_{(B^{\pi})^1}$, if $A^o = B^o = 0$,

   and an analogous statement for A^2, B^2 instead of A^1, B^1.

Proof:

1. Let $L \in S^o$ be i-previsible. Then by definition of S^o and theorem (4.2)

$$m_{A^o}(L) = m_A(L) = m_{A^{\pi_i}}(L) = m_{(A^{\pi_i})^o}(L).$$

If L is previsible, use theorem (10.3) instead of theorem (4.2).

2. Now assume $m_{A^o} \ll m_{B^o}$ and let again $L \in S^o$ be i-previsible. If $m_{(B^{\pi_i})^o}(L) = 0$, then by part 1 $m_{B^o}(L) = 0$. Hence $m_{A^o}(L) = 0$

and again part 1 gives $m_{(A^{\pi_i})^o}(L) = 0$. This yields part 3, since the previsible case is identical.

3. To prove part 2, assume $A^o = 0$. Then by part 3, $(A^{\pi_i})^o = 0$. Let $L \in s(S^1)$ be i-previsible. Again by the definition of $s(S^1)$ and theorem (4.2)

$$m_{A^1}(L) = m_A(L) = m_{A^{\pi_i}}(L) = m_{(A^{\pi_i})^1}(L).$$

In the previsible case, theorem (10.3) replaces theorem (4.2). This yields part 2.

4. Since $A^o = B^o = 0$, also $(A^{\pi_i})^o = (B^{\pi_i})^o = 0$, by part 1. Now argue like in 2. □

We obtain the following criterion for inaccessibility.

Proposition 2:

1. Let $L \in S^o$. Then:

   i) L is totally inaccessible iff $[\Gamma(L)^{\pi_1}]^o = 0 = [\Gamma(L)^{\pi_2}]^o$,

   ii) L is 1-previsible, 2-inaccessible iff $L \in \mathcal{P}^1$, $[\Gamma(L)^{\pi_2}]^o = 0$.

   An analogous statement holds for 2-previsible, 1-inaccessible 0-simple sets.

2. Let $L \in s(S^i)$. Then L is inaccessible iff $[\Gamma(L)^{\pi_i}]^i = 0$, $i=1,2$.

Proof:

1. First note that for $K \in S^o$ by definition of $\Gamma(L)$

$$P(\pi_\Omega(K \cap L)) = 0 \quad \text{iff} \quad m_{\Gamma(L)}(K) = 0.$$

Therefore, assertion 1 follows directly from part 2 of proposition 1 and theorem (12.1).

2. Let $L \in s(S^1)$. Then by definition $\Gamma(L)^o = \Gamma(L)^2 = 0$. Moreover, for $K \in S^1$ again

$$P(\pi_\Omega(K \cap L)) = 0 \quad \text{iff} \quad m_{\Gamma(L)}(K) = 0.$$

The conclusion is drawn from part 4 of proposition 1 and theorem (12.1). □

As a consequence of proposition 2, we note that for 0-simple sets the geometrical property characterizing i-previsibility, $\bar{i}$-inaccessibility can be made more strict.

Proposition 3: Let $L \in S^o$. Then:
L is 1-previsible, 2-inaccessible iff $L \in \mathcal{P}^1$ and for any 2-previsible $K \in S^o$

$P(\pi_\Omega(K \cap L)) = 0.$

An analogous statement holds for 2-previsible, 1-inaccessible sets.

Proof:
Let L be 1-previsible, 2-inaccessible and let $K \in S^o$ be 2-previsible. By proposition 2 we have $[\Gamma(L)^{\pi_2}]^o = 0$. Now by proposition 1, part 1

$$P(\pi_\Omega(K \cap L)) \leq m_{\Gamma(L)}(K) = m_{[\Gamma(L)^{\pi_2}]^o}(K) = 0.$$

This proves the non-trivial implication. □

We are prepared to state the main results of this section.

Theorem 1: Let $p \geq 1$, $L \in S^o$ p-integrable. Then there exist
i) a totally inaccessible $S \subset L$,
ii) a sequence $(T_n)_{n \in \mathbb{N}}$ of 1-previsible, 2-inaccessible sets,
iii) a sequence $(U_n)_{n \in \mathbb{N}}$ of 2-previsible, 1-inaccessible sets,
iv) a sequence $(V_n)_{n \in \mathbb{N}}$ of previsible sets in S^o,
such that
v) $L \subset S \cup \bigcup_{n \in \mathbb{N}} [T_n \cup U_n \cup V_n]$,
vi) the sets in i)-iv) are pairwise disjoint and p-integrable.
If $L \in \mathcal{P}^1$, the sets in i) and iii) can be chosen empty. An analogous statement holds for $L \in \mathcal{P}^2$.

Proof:
Let $(V_n)_{n \in \mathbb{N}}$ be a pairwise disjoint sequence of previsible sets in S^o decomposing $[\Gamma(L)^\pi]^o$ according to theorem (12.1), $(U_n')_{n \in \mathbb{N}}$ resp. $(T_n')_{n \in \mathbb{N}}$ corresponding sequences for $[\Gamma(L)^{\pi_2}]^o$ resp.

$[\Gamma(L)^{\pi_1}]^{\circ}$. Define for $m \in \mathbb{N}$

$$U_m = U'_m \setminus \bigcup_{n \in \mathbb{N}} V_n,$$

$$T_m = T'_m \setminus \bigcup_{n \in \mathbb{N}} V_n,$$

$$S = L \setminus \bigcup_{n \in \mathbb{N}} [U'_n \cup T'_n \cup V_n].$$

Clearly, iv) and v) are satisfied. Moreover, once we have shown that S is totally inaccessible, T_n 1-previsible, 2-inaccessible, U_n 2-previsible, 1-inaccessible for $n \in \mathbb{N}$, the definitions and proposition 3 will imply that the sets in i)-iv) are pairwise disjoint. For p-integrability consult theorem (12.1), theorem (4.3) and theorem (10.4). To prove that S is totally inaccessible, let $W = \bigcup_{n \in \mathbb{N}} U'_n$ and $A = \Gamma(S)$, $B = \Gamma(L \cap \bar{W})$. Then obviously $m_{A^{\circ}} \ll m_{B^{\circ}}$. Hence by proposition 1, part 3

$$m_{(A^{\pi_2})^{\circ}} \ll m_{(B^{\pi_2})^{\circ}}.$$

But, since $W \in \mathcal{P}^2$,

$$(B^{\pi_2})^{\circ} = [\Gamma(L \cap \bar{W})^{\pi_2}]^{\circ} = [\Gamma(L)^{\pi_2}(\bar{W})]^{\circ} = 0.$$

Therefore $(A^{\pi_2})^{\circ} = 0$. Analogously, $(A^{\pi_1})^{\circ} = 0$. By proposition 2, S is totally inaccessible.

Now fix $m \in \mathbb{N}$. In the arguments just given replace S by U_m, W by $\bigcup_{n \in \mathbb{N}} V_n$ and observe $U_m \in \mathcal{P}^2$. Then propositions 1 and 2 imply that U_m is 2-previsible, 1-inaccessible. A similar procedure works for T_m. The rest is obvious. □

<u>Corollary</u>: Let $L \in \mathcal{S}^{\circ}$. Then there exists a unique partition of L by sets $i(L)$, $i_1(L)$, $i_2(L)$, $a(L) \in \mathcal{S}^{\circ}$ such that for any family of 0-simple sets according to theorem 1

$$i(L) = S,\ i_1(L) = L \cap \bigcup_{n \in \mathbb{N}} U_n,\ i_2(L) = L \cap \bigcup_{n \in \mathbb{N}} T_n,\ a(L) = L \cap \bigcup_{n \in \mathbb{N}} V_n.$$

<u>Definition 2</u>: Let $L \in \mathcal{S}^{\circ}$. Then, in the notation of the preceding corollary, $i(L)$ is called "<u>totally inaccessible part</u>" of L,

$i_1(L)$ "<u>1-inaccessible, 2-accessible part</u>" of L, $i_2(L)$ "<u>2-inaccessible, 1-accessible part</u>" of L, $a(L)$ "<u>accessible part</u>" of L.

For the other two kinds of simple sets, we have the following results.

<u>Theorem 2</u>: Let $p \geq 1$, $L \in s(S^1)$ p-integrable. Then there exist

i) an inaccessible $S \subset L$,

ii) a sequence $(T_n)_{n \in \mathbb{N}}$ of previsible sets in $s(S^1)$,

such that

iii) $L \subset S \cup \bigcup_{n \in \mathbb{N}} T_n$,

iv) the sets in i), ii) are pairwise disjoint and p-integrable.

An analogous statement holds for S^1 replaced by S^2.

<u>Proof</u>:

Choose a pairwise disjoint sequence $(T_n)_{n \in \mathbb{N}}$ of sets in $s(S^1)$ decomposing $[\Gamma(L)^{\pi_1}]^1$ according to theorem (12.1) and observe that by proposition 1 we have $[\Gamma(L)^{\pi_1}]^o = [\Gamma(L)^{\pi_1}]^2 = 0$. Set

$$S = L \setminus \bigcup_{n \in \mathbb{N}} T_n.$$

Then ii), iii) and the first part of iv) are trivial. Theorems (12.1), (4.3) and (10.4) take care of the p-integrability assertion. To prove i), note that for any $\omega \in \Omega$ the set $\bigcup_{n \in \mathbb{N}} (T_n)_\omega$ is a countable union of open vertical line segments in Π whose upper boundaries are on $\partial\Pi$. Therefore $S \in s(S^1)$. The proof of the inaccessibility of S is easier than the corresponding part of the proof of theorem 1 and proceeds along the same lines of reasoning. □

<u>Corollary</u>: Let $L \in s(S^1)$. Then there exists a unique partition of L by sets $i(L)$, $a(L) \in s(S^1)$ such that for any family of sets according to theorem 2

$$i(L) = S, \quad a(L) = L \cap \bigcup_{n \in \mathbb{N}} T_n.$$

An analogous statement holds for S^1 replaced by S^2.

Definition 3: Let $i=1,2$, $L \in s(S^i)$. Then, in the notation of the preceding corollary, $i(L)$ is called "inaccessible part" of L, $a(L)$ "accessible part" of L.

Remark: We could have defined the different parts in the decomposition of simple sets according to the corollaries of theorems 1 and 2 in a more compact form. For example, for $L \in S^o$

$$a(L) = L \cap \{\Delta.\Gamma(L)^\pi > 0\}.$$

But in section 15 we will need an explicit description of $\{\Delta.\Gamma(L)^\pi > 0\}$ by simple sets. This mainly motivated the forms of the theorems of this section.

14. The jumps of regular martingales

To prepare the ground for the structure theorem for square integrable martingales in section 19, we return to martingale theory now. In section 9 we learned that any $L\log^+L$-bounded martingale possesses a regular version. The results of section 1 suggest that the set of 0-jumps of a regular martingale is located on a countable union of 0-simple sets, the set of 1- and 2-jumps of a regular martingale without 0-jumps on a countable union of sets in $s(S^1)$ and $s(S^2)$. However, the results we need in section 19 require a closer look at the decomposition of the sets of discontinuities. The technique we apply to decompose square integrable martingales consists in gradually removing their jumps, beginning with the 0-jumps. Let us briefly concentrate on them and assume $(L_n)_{n \in \mathbb{N}}$ is a sequence of 0-simple sets containing all 0-jumps of M. We try to find a martingale M^o which contains all these jumps. Let $M(L_n)$ be the "jump of M" on L_n, considered as a process of bounded variation. To make M^o a martingale, we have to "compensate" $M(L_n)$ by a process A_n such that $M(L_n)-A_n$ is a martingale. On the other hand, to make it contain all 0-jumps, A_n is not allowed to possess discontinuities of this kind. As we will see in the next section, all these requirements can be met if we decompose the set of 0-jumps of M in advance by 0-simple sets of the different accessibility degrees discussed in the preceding section. This will be done in the main theorems.

We start by defining "jumps" of regular martingales M on i-simple sets L, i=0,1,2. For i=0, M(L) is the integral process of $\Delta.M$ with respect to $\Gamma(L)$. If i=1, it is an analogue of the integral defining $\Gamma(L)$ with the stochastic measure given by $t_2 \to \Delta^1_{t_1} M_{(\cdot,t_2)}$ replacing the Lebesgue measure on the vertical line segments constituting L. At this point, we realize a discrepancy with the one-parameter theory which has already been mentioned: jumps of one-parameter martingales on graphs of stopping times inherit the integrability properties of the martingales; jumps of two-parameter martingales do not, since the II-sections of simple sets are only finite. We therefore have to derive some simple integrability results together with the basic properties of the jump processes of martingales. The main theorems describe the decompositions of the discontinuity sets of $L\log^+L$-bounded martingales in terms of simple sets of the kinds considered in section 13. Mazziotto, Szpirglas [26] show that the jumps of regular processes are located on "simple" stopping lines.

We define jumps on 1- or 2-simple sets only for martingales without 0-jumps, since for investigating them we will always remove all the 0-jumps in advance.

Definition 1:

1. Let M be a regular martingale, $L \in S^{o}$. Then

$M(L) = \Delta.M \cdot \Gamma(L).$

is called "0-jump of M on L".

2. Let M be a regular martingale without 0-jumps, $L \in S^{1}$. Then

$$M(L)_t = \int_{[0,t]} (\Delta^1_{s_1} M_{(.,t_2)} - \Delta^1_{s_1} M_{(.,s_2)})\, d\Gamma(\partial L)_s \quad , \quad t \in \mathbb{I} ,$$

is called "1-jump of M on L".

For $L \in s(S^1)$, M(L) is defined by additivity.

In a similar way, a "2-jump of M on L" is defined for $L \in s(S^2)$.

Remarks: 1. If M is a regular martingale and $L \in S^{o}$, we have $M(L) \in \mathfrak{m}(\mathfrak{G}, \mathfrak{B}(\mathbb{R}))$ since according to theorem (10.1), $\Delta.M$ is optional.

2. If M is a regular martingale without 0-jumps and $L \in s(S^1)$, we have $M(L) \in \mathfrak{m}(\mathfrak{G}^1 \cap \mathfrak{P}^2, \mathfrak{B}(\mathbb{R}))$ since $\Delta^1.M_{(.,.)}$ is optional by theorem (10.1) and continuous in the second parameter.

3. If $L \in \mathfrak{P}$, definition 1 and definition (9.3) of M(L) are consistent. This will become clear in section 15.

Given a regular martingale M and a simple set L, the jump M(L) is not necessarily integrable. We therefore note the following integrability properties.

Proposition 1: Let $p>1$, $q,r \geq 1$ such that $\frac{1}{r} = \frac{1}{p} + \frac{1}{q}$, M a regular martingale. Then: if M is p-integrable and $L \in S^{o}$ q-integrable,

$\overline{M}(L) = (\Delta.M)^{+} \cdot \Gamma(L).$, $\underline{M}(L) = (\Delta.M)^{-} \cdot \Gamma(L).$

are r-integrable optional increasing processes.

In particular, M(L) is a process of r-integrable variation.

Proof:

Optionality is clear from remark 1 after definition 1. Let us argue for $\overline{M}(L)$. Clearly, $\overline{M}(L)$ is an increasing process which satisfies

$$\overline{M}(L)_{(1,1)} \le 4 \sup_{t\in\Pi} |M_t| \; \Gamma(L)_{(1,1)}.$$

Now $\Gamma(L)_{(1,1)}$ is q-integrable by hypothesis, whereas $\sup_{t\in\Pi} |M_t|$ is p-integrable by theorem (8.1). Hölder's inequality yields r-integrability of $\overline{M}(L)_{(1,1)}$. Finally, $M(L) = \overline{M}(L) - \underline{M}(L)$. This completes the proof. □

By definition as a partial stochastic integral, jumps on 1- or 2-simple sets also possess some martingale type properties.

Proposition 2: Let $p>1$, $q,r\ge 1$ such that $\frac{1}{r} = \frac{1}{p} + \frac{1}{q}$, M a regular martingale without 0-jumps. Then:

if M is p-integrable and $L \in \mathcal{S}^1$ is q-integrable,

$$\overline{M}(L)_t = \int_{[0,t]} (\Delta^1_{s_1} M_{(\cdot,t_2)} - \Delta^1_{s_1} M_{(\cdot,s_2)})^+ \, d\Gamma(\partial L)_s,$$

$$\underline{M}(L)_t = \int_{[0,t]} (\Delta^1_{s_1} M_{(\cdot,t_2)} - \Delta^1_{s_1} M_{(\cdot,s_2)})^- \, d\Gamma(\partial L)_s, \quad t\in\Pi,$$

are r-integrable 2-submartingales of class D, D^1 and D^2 which are $\hat{Q}^{++}$-continuous in $L^1(\Omega,\mathcal{F},P)$, increasing in the first and continuous in the second parameter.

Moreover, M(L) is an r-integrable 2-martingale which is continuous in the second parameter.

An analogous result holds for $L \in s(\mathcal{S}^1)$ and with $\mathcal{S}^2$ instead of $\mathcal{S}^1$.

Proof:

Let $L \in \mathcal{S}^1$. We argue for $\overline{M}(L)$. It is clear that $\overline{M}(L)$ is increasing in the first parameter. Since M has no 0-jumps, it is continuous in the second parameter. The r-integrability assertion follows like in the proof of proposition 1, since we have

$$\overline{M}(L)_{(1,1)} \le 4 \sup_{t\in\Pi} |M_t| \; \Gamma(\partial L)_{(1,1)}.$$

Since M is regular, $\overline{M}(L)$ is $\hat{Q}^{++}$-continuous in $L^1(\Omega,\mathcal{F},P)$. The 2-martingale property of M(L) is an easy consequence of the martingale property of M and adaptedness of $\Gamma(\partial L)$. To prove that $\overline{M}(L)$ is a 2-submartingale, let $J =]u,v] \in \mathcal{J}$ be given. Then

$$\Delta_J\overline{M}(L) = \Delta_{J_2} \int_{J_1\times[0,.]} (\Delta^1_{s_1} M_{(.,.)} - \Delta^1_{s_1} M_{(.,s_2)})^+ \, d\Gamma(\partial L)_s.$$

Furthermore, by the martingale property of M

$$E\Big(\int_{J_1\times[0,v_2]} (\Delta^1_{s_1} M_{(.,v_2)} - \Delta^1_{s_1} M_{(.,s_2)})^+ \, d\Gamma(\partial L)_s \,|\, \mathcal{F}^2_{u_2}\Big)$$

$$\geq \int_{J_1\times[0,u_2]} E((\Delta^1_{s_1} M_{(.,v_2)} - \Delta^1_{s_1} M_{(.,s_2)})^+ |\mathcal{F}^2_{u_2}) \, d\Gamma(\partial L)_s$$

$$\geq \int_{J_1\times[0,u_2]} (\Delta^1_{s_1} M_{(.,u_2)} - \Delta^1_{s_1} M_{(.,s_2)})^+ \, d\Gamma(\partial L)_s.$$

This, however, implies

$$E(\Delta_J\overline{M}(L)|\mathcal{F}^2_{u_2}) \geq 0,$$

which means that $\overline{M}(L)$ is a 2-submartingale.

According to Métivier [31], p. 96 and p. 53, the non-negative submartingale $\overline{M}(L)_{(1,.)}$ possesses a Doléans measure. Therefore, $\overline{M}(L)$ is of class D^2. It is also of class D^1 and of class D, since it is increasing in the first variable (see the proof of proposition (8.4), where theorem (4.3) is used). The extension of these results to $s(\mathcal{S}^1)$ poses no new problems. □

The 0-jumps of an L $\log^+$L-bounded martingale can be decomposed in the following way.

<u>Theorem 1</u>: Let M be an L $\log^+$L-bounded regular martingale. Then there exist

i) a sequence $(S_n)_{n\in\mathbb{N}}$ of totally inaccessible sets,

ii) a sequence $(T_n)_{n\in\mathbb{N}}$ of 1-previsible, 2-inaccessible sets,

iii) a sequence $(U_n)_{n\in\mathbb{N}}$ of 2-previsible, 1-inaccessible sets,

iv) a sequence $(V_n)_{n\in\mathbb{N}}$ of previsible sets in $\mathcal{S}^o$,

such that

v) M has no 0-jumps outside $\bigcup_{n\in\mathbb{N}} [S_n \cup T_n \cup U_n \cup V_n]$,

vi) the sets in i)-iv) are pairwise disjoint and p-integrable for any $p\geq 0$.

If $\Delta.M \in \mathfrak{M}(\mathcal{P}^1, \mathcal{B}(\mathbb{R}))$, the sets in i) and iii) can be chosen empty.

An analogous statement holds for $\mathfrak{P}^2$.

Proof:

1. For $n \in \mathbb{N}$, let M^n be a regular version of the bounded martingale

$$E(-n \vee (M_{(1,1)} \wedge n) \mid \mathcal{F}.).$$

By theorem (8.1), the set of 0-jumps of M is contained in the countable union of the sets of 0-jumps of M^n, $n \in \mathbb{N}$. It is therefore enough to prove the theorem for bounded M.

2. Assume M is bounded. For $\varepsilon \geq 0$, let

$$L^O(\varepsilon) = \{(\omega,t) \in \Omega \times \mathbb{I} : |\Delta_t M(\omega)| > \varepsilon\}.$$

Fix $\varepsilon > 0$. We know that $\Delta.M$ is optional by theorem (10.1). Therefore $L^O(\varepsilon) \in \mathcal{G}$. Moreover, theorem (1.1) tells us that $L^O(\varepsilon)_\omega$ is finite for any $\omega \in \Omega$. Therefore for any 0-sequence $(\mathbb{J}_n)_{n \in \mathbb{N}}$ of partitions of $\mathbb{I}$, any $\omega \in \Omega$

$$|L^O(\varepsilon)_\omega| \leq \frac{1}{\varepsilon^2} \liminf_{n\to\infty} \sum_{J \in \mathbb{J}_n} (\Delta_J M)^2(\omega).$$

Fatou's lemma and theorem (8.2) imply that $L^O(\varepsilon) \in \mathcal{S}^O$ is p-integrable for any $p \geq 0$. Now let

$$K_n = L^O(\tfrac{1}{n}), \quad n \in \mathbb{N}.$$

Then $(K_n)_{n \in \mathbb{N}}$ is a sequence of p-integrable sets in $\mathcal{S}^O$ which cover $L^O(0)$, $p \geq 0$.

3. We finally apply the process of decomposition presented in theorem (13.1) to every K_n, $n \in \mathbb{N}$. This produces

i') a sequence $(S_n)_{n \in \mathbb{N}}$ of totally inaccessible sets,

ii') sequences $(T_{n,m})_{n,m \in \mathbb{N}}$ of 1-previsible, 2-inaccessible sets,

iii') sequences $(U_{n,m})_{n,m \in \mathbb{N}}$ of 2-previsible, 1-inaccessible sets,

iv') sequences $(V_{n,m})_{n,m \in \mathbb{N}}$ of previsible sets in $\mathcal{S}^O$,

such that

v') $L^O(0) = \bigcup_{n \in \mathbb{N}} K_n \subset \bigcup_{n,m \in \mathbb{N}} [S_n \cup T_{n,m} \cup U_{n,m} \cup V_{n,m}]$,

vi') the sets in i')-iv') are p-integrable for all $p \geq 0$.

It remains to rearrange the double sequences and make them pairwise disjoint in the usual way. The previsibility assertion is a consequence of the corresponding statement in theorem (13.1).□

For the 1- and 2-jumps of an $L\log^+L$-bounded martingale without 0-jumps we have the following result.

<u>Theorem 2</u>: Let M be an $L\log^+L$-bounded regular martingale which has no 0-jumps. Then there exist

i) a sequence $(S_n)_{n\in\mathbb{N}}$ of inaccessible sets in $s(S^1)$,

ii) a sequence $(T_n)_{n\in\mathbb{N}}$ of previsible sets in $s(S^1)$,

iii) a sequence $(U_n)_{n\in\mathbb{N}}$ of inaccessible sets in $s(S^2)$,

iv) a sequence $(V_n)_{n\in\mathbb{N}}$ of previsible sets in $s(S^2)$,

such that

v) M is continuous outside $\bigcup_{n\in\mathbb{N}} [S_n \cup T_n \cup U_n \cup V_n]$,

vi) the sets in i), ii) resp. iii), iv) are pairwise disjoint and p-integrable for any $p\geq 0$.

<u>Proof</u>:

1. Like in the proof of theorem 1, it is enough to consider a bounded martingale M. Let us concentrate on the 1-jumps. For $\varepsilon\geq 0$, set

$$L^1(\varepsilon) = \{(\omega,t)\in\Omega\times\mathrm{I\!I} : \quad |\Delta^1_{t_1} M_{(.,t_2)}(\omega)| > \varepsilon\}.$$

Fix $\varepsilon>0$. The process $\Delta^1_{.}M_{(.,.)}$ is $\mathfrak{G}$-measurable by theorem (10.1) and continuous in the second variable, since M has no 0-jumps. Therefore, $L^1(\varepsilon)\in\mathfrak{G}^1\cap\mathfrak{P}^2$. By theorem (1.2), $L^1(\varepsilon)_\omega$ is contained in a finite union of open vertical line segments and consists of countably many open vertical line segments for $\omega\in\Omega$. Now let

$$G(\varepsilon) = \bigcup_{q\in\mathbb{Q}_+} [L^1(\varepsilon) + \Omega\times\{(0,q)\}]\cap(\Omega\times\mathrm{I\!I}),$$

$H(\varepsilon)$ the random set of boundary points of the line segments in $G(\varepsilon)$.

We will show $H(\varepsilon)\in S^0$. We already know that $G(\varepsilon)\in\mathfrak{G}$. By the proof of proposition (12.2), which does not involve any integrability assumption, $H(\varepsilon) \in \mathbb{G}$ and therefore

$$A(\varepsilon)_t = |H(\varepsilon) \cap [0,t]|, \quad t \in \mathbb{I},$$

is an optional increasing process. By continuity for any $\omega \in \Omega$

$$|H(\varepsilon)_\omega| \le \frac{1}{\varepsilon^2} \int_{\mathbb{I}} (\Delta^1_{s_1} M_{(.,s_2)})^2 \, dA(\varepsilon)_s \, (\omega).$$

Therefore, for any 0-sequence $(\mathbb{J}_n)_{n \in \mathbb{N}}$ of partitions of $[0,1]$, any $p \ge 1$

$$\| H(\varepsilon) \|_p \le \frac{1}{\varepsilon^2} \| \int_{\mathbb{I}} (\Delta^1_{s_1} M_{(.,s_2)})^2 \, dA(\varepsilon)_s \|_p$$

$$\le \frac{1}{\varepsilon^2} \| \int_{\mathbb{I}} (\Delta^1_{s_1} M_{(.,1)})^2 \, dA(\varepsilon)_{(s_1,1)} \|_p \quad \text{(M is a martingale)}$$

$$\le \frac{1}{\varepsilon^2} \| \liminf_{n\to\infty} \sum_{J \in \mathbb{J}_n} (\Delta_J M_{(.,1)})^2 \|_p$$

$$\le \frac{1}{\varepsilon^2} \liminf_{n\to\infty} \| \sum_{J \in \mathbb{J}_n} (\Delta_J M_{(.,1)})^2 \|_p \quad \text{(Fatou's lemma).}$$

Theorem (8.2) implies that $H(\varepsilon) \in \mathcal{S}^o$ is p-integrable for any $p \ge 0$. Finally,

$$L^1(\varepsilon) \subset G(\varepsilon) = \sigma_1 H(\varepsilon) \in \mathcal{S}^1$$

and $G(\varepsilon)$ is p-integrable for any $p \ge 0$. Now let

$$K_n = G(\tfrac{1}{n}), \quad n \in \mathbb{N}.$$

Then $(K_n)_{n \in \mathbb{N}}$ is a sequence of p-integrable sets in $\mathcal{S}^1$ which cover $L^1(0)$, $p \ge 0$.

2. We now apply theorem (13.2) to every K_n, $n \in \mathbb{N}$. This yields

i') a sequence $(S_n)_{n \in \mathbb{N}}$ of inaccessible sets in $s(\mathcal{S}^1)$,

ii') sequences $(T_{n,m})_{n,m \in \mathbb{N}}$ of previsible sets in $s(\mathcal{S}^1)$,

such that

iii') $L^1(0) = \bigcup_{n \in \mathbb{N}} K_n \subset \bigcup_{n,m \in \mathbb{N}} [S_n \cup T_{n,m}]$,

iv') the sets in i'), ii') are p-integrable for any $p \ge 0$.

Complete the proof by rearranging and making the sequences pairwise disjoint. □

15. The compensation of martingale jumps

Let M(L) be the jump of a regular martingale on a simple set L which is of one of the types described in the decomposition theorems (14.1) and (14.2). As indicated in the beginning of the last section we now have to find a compensator of M(L), i.e. a process C which satisfies the requirements

i) M(L) - C is a martingale,

ii) C is of a "higher degree of continuity" than M(L).

By ii) we mean that if $L \in S^o$, C should have only 1- or 2-jumps and if $L \in s(S^i)$, i=1,2, C should be continuous. This condition will imply that the successive extraction of a "0-jump part", "1-jump part" and "2-jump part" will indeed improve the continuity properties of the respective remainders.

To obtain a process C satisfying i) we proceed in the following way. In consequence of the results of propositions (14.1) and (14.2), for appropriately chosen integrability degrees of M and L, $\underline{M}(L)$ and $\overline{M}(L)$ are integrable increasing processes resp. submartingales of class D, D^1 and D^2. Therefore, the decomposition theorems of sections 5 and 11 will yield a process C which fulfills i). Now L already possesses the properties which will automatically guarantee ii) for such a C. Roughly speaking, the reason is this: if L is previsible in i-direction, M(L) behaves like a martingale in this direction and need not be compensated; if, however, L is inaccessible in i-direction, the compensator C, being a combination of dual previsible projections, cannot see the jump and smoothes it out, i.e. it behaves continuously in this direction.

We first prove that, if L is i-previsible, M(L) is an i-martingale. After obtaining the compensators, we investigate their smoothness properties. If $L \in S^o$, the methods of section 13 are applicable, since $\overline{M}(L)$, $\underline{M}(L)$ are absolutely continuous with respect to $\Gamma(L)$. For $L \in s(S^i)$, i=1,2, we study directly its various jump sets.

Proposition 1: Let $M \in M^2$, $L \in S^o$ 2-integrable, i=1,2. Then M(L) is an i-martingale, if $L \in \mathbf{P}^i$. Moreover, M(L) is a martingale, if $L \in \mathbf{P}$.

Proof:

By uniform approximation using theorem (8.1) and proposition (14.1) we may assume M to be bounded. Let i=1 and $L \in \mathbf{P}^1$. Theorem (12.2) gives $\Gamma(L) \in \mathfrak{M}(\mathbf{P}^1, \mathbf{B}(\mathbb{R}))$. Now pick $F \times J \in \mathbf{R}^1$. Using theorems (4.1) and (10.1) we obtain

$$
\begin{aligned}
E(1_F \, \Delta_J M(L)) &= \int_{\Omega \times \Pi} 1_{F \times J} \, \Delta.M \, dm_{\Gamma(L)} \\
&= \int_{\Omega \times \Pi} 1_{F \times J} \, {}^{\pi_1}\Delta.M \, dm_{\Gamma(L)} \\
&= \int_{\Omega \times \Pi} 1_{F \times J} \, [M^{-\cdot} - M^{-\cdot} - M^{--} + M^{--}] \, dm_{\Gamma(L)} \\
&= 0.
\end{aligned}
$$

This means that M(L) is a 1-martingale.
If $L \in \mathbf{P}$, M(L) is both a 1- and a 2-martingale, hence a martingale. □

For 1-jumps we already know that M(L) is a 2-martingale. Therefore only 1-previsibility of L matters.

Proposition 2: Let $M \in M^2$, i=1,2, $L \in s(S^i)$ 2-integrable. Then M(L) is a martingale, if $L \in \mathbf{P}$.

Proof:

By a uniform approximation argument using theorem (8.1) and proposition (14.1) we may consider M to be bounded. Take i=1 and let $L \in \mathcal{P}$. Additivity allows us to assume $L \in S^1$. By theorem (12.2) and proposition (12.2), $\Gamma(\partial L) \in \mathfrak{m}(\mathcal{P}^1, \mathcal{B}(\mathbb{R}))$. Let $F \times J \in \mathcal{R}^1$. For simplicity, suppose $J =]s_1,t_1] \times [0,t_2]$. Using theorems (4.1) and (10.1) again, we obtain

$$E(1_F \, \Delta_J M(L)) = \int_{\Omega \times \Pi} 1_{F\times J}[\Delta^1_{s_1} M_{(.,t_2)} - \Delta^1_{s_1} M_{(.,s_2)}] \, dm_{\Gamma(\partial L)}(s,.)$$

$$= \int_{\Omega \times \Pi} 1_{F\times J} [{}^{\pi_1}\Delta^1_{s_1} M_{(.,t_2)} - {}^{\pi_1}\Delta^1_{s_1} M_{(.,s_2)}] \, dm_{\Gamma(\partial L)}(s,.)$$

$$= \int_{\Omega \times \Pi} 1_{F\times J} [M^{-\cdot}_{(s_1,t_2)} - M^{-\cdot}_{(s_1,t_2)} - M^{-\cdot}_{s} + M^{-\cdot}_{s}] \, dm_{\Gamma(\partial L)}(s,.)$$

$$= 0.$$

This expresses that $M(L)$ is a 1-martingale. Therefore, $M(L)$ is a martingale. □

Let us now study those time directions in which a given simple set L is inaccessible and compute the compensator of $M(L)$. We start with totally inaccessible L.

<u>Theorem 1</u>: Let $M \in M^2$, $L \in S^o$ 2-integrable and totally inaccessible, $\overline{M}(L)$, $\underline{M}(L)$ like in proposition (14.1). Then the process

$$C = [\overline{M}(L)^{\pi_1} + \overline{M}(L)^{\pi_2} - \overline{M}(L)^{\pi}] - [\underline{M}(L)^{\pi_1} + \underline{M}(L)^{\pi_2} - \underline{M}(L)^{\pi}]$$

satisfies

i) $M(L) - C$ is a martingale,

ii) C possesses no 0-jumps.

<u>Proof</u>:

According to corollary 1 of theorem (11.2), we only need to prove that the 0-parts of $\overline{M}(L)^{\pi_1}, \ldots, \underline{M}(L)^{\pi}$ vanish. Now observe that in virtue of the definition of $\overline{M}(L)$, $\underline{M}(L)$ we have

$$m_{\overline{M}(L)^o} \ll m_{\Gamma(L)^o}, \quad m_{\underline{M}(L)^o} \ll m_{\Gamma(L)^o}.$$

Therefore, by proposition (13.1)

$$m_{(\overline{M}(L)^{\alpha})^o} \ll m_{(\Gamma(L)^{\alpha})^o}, \quad m_{(\underline{M}(L)^{\alpha})^o} \ll m_{(\Gamma(L)^{\alpha})^o} \quad \text{for } \alpha = \pi_1, \pi_2, \pi.$$

But, by proposition (13.2), $(\Gamma(L)^{\pi_1})^o = (\Gamma(L)^{\pi_2})^o = 0$. Hence also $(\Gamma(L)^{\pi})^{o} = 0$, again by proposition (13.1). This completes the proof. □

Next, we consider "partially" inaccessible 0-simple sets.

Theorem 2: Let $M \in \mathcal{M}^2$, $L \in \mathcal{S}^o$ 2-integrable and 1-previsible, 2-inaccessible, $\overline{M}(L)$, $\underline{M}(L)$ like in proposition (14.1). Then the process

$$C = \overline{M}(L)^{\pi_2} - \underline{M}(L)^{\pi_2}$$

satisfies

i) $M(L) - C$ is a martingale,

ii) C possesses no 0-jumps.

An analogous statement holds for 2-previsible, 1-inaccessible sets.

Proof:

Since $M(L)$ is a 1-martingale, as follows from proposition 1, we have

$$m_{\overline{M}(L)}|\mathcal{P}^1 = m_{\underline{M}(L)}|\mathcal{P}^1.$$

Therefore $\overline{M}(L)^{\pi_1} = \underline{M}(L)^{\pi_1}$ and consequently $\overline{M}(L)^{\pi} = \underline{M}(L)^{\pi}$. Corollary 1 of theorem (11.2) now says that all we need to show is that the 0-parts of $\overline{M}(L)^{\pi_2}$, $\underline{M}(L)^{\pi_2}$ vanish. To do this, we make use again of the results of section 13. First observe

$$m_{\overline{M}(L)^o} \ll m_{\Gamma(L)^o}, \quad m_{\underline{M}(L)^o} \ll m_{\Gamma(L)^o}$$

as in the preceding proof. Therefore

$$m_{(\overline{M}(L)^{\pi_2})^o} \ll m_{(\Gamma(L)^{\pi_2})^o}, \quad m_{(\underline{M}(L)^{\pi_2})^o} \ll m_{(\Gamma(L)^{\pi_2})^o}.$$

Now proposition (13.2) implies $(\Gamma(L)^{\pi_2})^o = 0$. This finishes the proof. □

The final theorem of this section deals with the 1- and 2-jumps of martingales which possess no 0-jumps. In view of proposition 2 we only have to consider inaccessible sets in $s(\mathcal{S}^i)$, i=1,2. We have to require a slightly sharper integrability assumption for the jump set. But in consideration of the results of section 14 this will not matter.

Theorem 3: Let $M \in \mathcal{M}^2$ be without 0-jumps, $L \in s(\mathcal{S}^1)$ p-integrable for some $p>2$ and inaccessible, $\overline{M}(L)$, $\underline{M}(L)$ like in proposition (14.2). Denote by $\overline{A}^1$, $\underline{A}^1$ the dual 1-previsible projections of $\overline{M}(L)$, $\underline{M}(L)$ according to theorem (5.2). Then a version of the process

$$C = \overline{A}^1 - \underline{A}^1$$

satisfies

i) $M(L) - C$ is a martingale,

ii) C is continuous.

An analogous statement holds for $L \in s(\mathcal{S}^2)$.

Proof:

We may assume $L \in \mathcal{S}^1$.

1. First note that, besides being 2-submartingales, $\overline{M}(L)$ and $\underline{M}(L)$ are also submartingales and weak submartingales of class D, D^1 and D^2. It suffices to remember that they are increasing in direction 1. Now theorem (5.1) produces unique integrable increasing processes $\overline{A}^2$, $\underline{A}^2 \in \mathfrak{m}(\mathcal{P}^2, \mathcal{B}(\mathbb{R}))$ such that $\overline{M}(L) - \overline{A}^2$, $\underline{M}(L) - \underline{A}^2$ are 2-martingales. Since $M(L)$ is a 2-martingale, we obtain $\overline{A}^2 = \underline{A}^2$. Now by theorem (11.2) also $\overline{A} = \underline{A}$, where $\overline{A}$, $\underline{A}$ are the dual previsible projections of $\overline{M}(L)$, $\underline{M}(L)$. By theorem (5.2), $\overline{A}^1$, $\underline{A}^1$ are well defined and theorem (11.2) again shows i).

2. The processes $\overline{M}(L)$, $\underline{M}(L)$ are r-integrable for some $r>1$ by proposition (14.2), hence C by theorem (5.3). Therefore theorem (9.1) implies that we can assume C to be regular and $\mathcal{P}^1$-measurable. Let us consider the regular martingale $N = M(L) - C$. Since M has no 0-jumps, we know that

$$\Delta.N = -\Delta.C \in \mathfrak{m}(\mathcal{P}^1, \mathcal{B}(\mathbb{R})).$$

If N is bounded, this implies

$$\Delta.N = {}^{\pi 1}\Delta.N = 0$$

according to theorem (10.1). If N is not necessarily bounded, consider this equation on the 1-previsible sets $S_n = \{\Delta.N \leq n\}$, $n \in \mathbb{N}$, to derive the same conclusion. We therefore obtain $\Delta.C = 0$, i.e. C has no 0-jumps.

Let us turn to the 1-jumps. Denote by $\Delta^1 N$ the process $t \to \Delta^1_{t_1} N_{(.,t_2)}$ and correspondingly $\Delta^1 C$. Then regularity, theorem (2.1) and continuity in the second parameter imply that $\Delta^1 C \in \mathfrak{m}(\mathcal{P}, \mathcal{B}(\mathbb{R}))$. Therefore $T = \{\Delta^1 C \neq 0\} \in \mathcal{P}$ and by definition of inaccessibility,

$$P(\pi_\Omega(T \cap L)) = 0.$$

Hence

$$1_T\, \Delta^1 N = -1_T\, \Delta^1 C \in \mathfrak{m}(\mathcal{P}, \mathcal{B}(\mathbb{R})).$$

If N is bounded, this entails again

$$1_T\, \Delta^1 N = 1_T\, {}^{\pi 1}\Delta^1 N = 0$$

according to theorem (10.1). If this is not the case, approximate on previsible sets as above. To summarize, $\Delta^1 C = 0$ and C is seen to have no 1-jumps as well.

We finally consider the 2-jumps. Let $K \in s(\mathcal{S}^2)$ be a component of the decomposition of the 2-jumps of N according to theorem (14.2). Then N(K) is well defined by proposition (14.2). It is a regular 1-martingale which is continuous in 1-direction. Moreover, due to the properties of $\bar{A}^1$, $\underline{A}^1$, the process $N(K)_{(.,t_2)}$ is of bounded variation for any $t_2 \in [0,1]$. Hence it vanishes by Dellacherie [16], p. 111. This implies $N(K) = 0$ and, since K was arbitrary and M(L) has no 2-jumps, that C has no 2-jumps. The proof of the continuity of C is complete. □

<u>Definition 1</u>: Let $M \in \mathcal{M}^2$, $i=0,1,2$, $L \in s(S^i)$ 2-integrable resp. p-integrable for some $p>2$. A process C according to theorems 1-3 is called "<u>compensator</u>" of M(L).

<u>Remark</u>: In case $L \in \mathcal{P}$, the process $C = 0$ is understood to be the compensator of M(L). This agrees with the results of propositions 1 and 2.

IV. Quadratic variation and structure of martingales

Compensated jumps of martingales on simple sets may be considered as their "elementary components" which can help to gain insight into the structure of martingales and their quadratic variations. To describe a way in which this can be done is our main occupation in this chapter. Section 16 provides some elementary general results about quadratic variations in L^p, which are defined as limits of square sums along a 0-sequence of partitions of Π for $p \geq 0$. In sections 17 and 18, before putting the "puzzle" together, we study its pieces more closely. In section 17, the quadratic variations of compensated i-jumps, $i=0,1,2$, and continuous martingales are consecutively computed. An important fact is investigated in section 18. Given two compensated jump parts of different type or with disjoint jump sets, we see that they do not "overlap": they have orthogonal variation, consequently are orthogonal in $\mathcal{M}^2$ and their quadratic variations simply add. Therefore, they also act independently as elementary components of a square integrable martingale. On the basis of this knowledge, our first main theorem is proved in section 19. It states that the orthogonal sum in $\mathcal{M}^2$ of the compensated 0-jumps of a square integrable martingale M belonging to a sequence of pairwise disjoint 0-simple sets which cover all 0-jumps of M defines a unique "0-jump part" M^o. Using sequences of pairwise disjoint sets in $s(S^i)$ which cover all i-jumps of $M-M^o$, "i-jump parts" M^i are uniquely determined, $i=1,2$. The martingale $M^c = M-M^o-M^1-M^2$ is continuous and the decomposition of M into these 4 parts is orthogonal in $\mathcal{M}^2$. Concerning quadratic variations, our second main theorem in section

20 finds a similar situation. The quadratic variations of M^i are simply the orthogonal sums of the quadratic variations of a sequence of corresponding elementary components. M^c possesses the quadratic variation $\langle M^c \rangle$. Hence M has a quadratic variation which is simply the sum of those of its four components. This detailed description is possible in M^2 where our structure theorem holds. For the larger class of $L\log^+L$-bounded martingales, using approximation, we still can prove that quadratic variation exists. But $L\log^+L$ is the boundary. In section 21, we construct an L^1-bounded martingale which has no quadratic variation.

16. Some general results on quadratic variation

In this introductory section we define quadratic variation and quadratic i-variation for an arbitrary process X and prove that the latter is an $\bar{i}$-submartingale if X is a square integrable $\bar{i}$-martingale. We then turn to some auxiliary results. The first group of them will be seen to have, roughly speaking, the consequence that compensation does not change the quadratic variation of a jump process. More precisely, all compensators appearing in section 15 will prove to have quadratic variation zero. We consider successively increasing processes without 0-jumps and continuous i-martingales which are increasing in $\bar{i}$-direction, i=1,2. Another group of auxiliary results is devoted to the question: Given a sequence $(M^n)_{n\in\mathbb{N}}$ of martingales which converge in some sense to a martingale M and which possess quadratic variations $[M^n]$, what can be said about the quadratic variation $[M]$ of M and convergence of $[M^n]$ to $[M]$?

Definition 1: Let $p \geq 0$, X a process on $\Omega \times \mathbb{I}$.

1. X is said to "possess a quadratic variation in L^p", if for any 0-sequence $(\mathcal{J}_n)_{n \in \mathbb{N}}$ of partitions of $\mathbb{I}$, any $t \in \mathbb{I}$, the sequence $\left(\sum_{J \in \mathcal{J}_n} (\Delta_{J \cap [0,t]} X)^2 \right)_{n \in \mathbb{N}}$ converges in $L^p(\Omega, \mathcal{F}, P)$. In this case, the L^p-limits determine an increasing process $[X]$ which is called "quadratic variation of X in L^p".

2. X is said to "possess a quadratic 1-variation in L^p", if for any 0-sequence $(\mathcal{J}_n)_{n \in \mathbb{N}}$ of partitions of $[0,1]$, any $t \in \mathbb{I}$, the sequence $\left(\sum_{J \in \mathcal{J}_n} (\Delta_{J \cap [0,t_1] \times [0,t_2]} X)^2 \right)_{n \in \mathbb{N}}$ converges in $L^p(\Omega, \mathcal{F}, P)$. In this case, the L^p-limits determine a process $[X]^1$ increasing in the first parameter which is called "quadratic 1-variation of X in L^p". In an analogous way, a "quadratic 2-variation of X in L^p" is defined.

If X is a square integrable martingale, $[X]^i$ possesses martingale type properties like $\langle X \rangle^i$, $i=1,2$, as the following proposition shows.

Proposition 1: Let M be a square integrable 2-martingale which possesses a quadratic 1-variation in L^1. Then $[M]^1$ is a 2-submartingale.
Moreover, for any $r \in [0,1]$, $([M]^1_{(r,.)})^{1/2}$ is a one-parameter submartingale. A corresponding statement holds for 1-martingales.

Proof:
Let $\mathcal{J}$ be a partition of a left open, right closed interval K_1 in $[0,1]$. Then

$$t_2 \to \left[\sum_{J \in \mathcal{J}} (\Delta_{J \times [0,t_2]} M)^2 \right]^{1/2}$$

is a submartingale by convexity of the 2-norm in finite dimensional euclidean spaces. Since M possesses a quadratic 1-variation in L^1, one can approximate $\Delta_{K_1} [M]^1_{(.,.)}$ by submartingales of this form in L^1. This gives the desired result. □

Let us now consider some processes whose quadratic variation is zero.

Proposition 2: Let $p\geq 0$, A a p-integrable increasing process without 0-jumps. Then A has quadratic variation 0 in $L^{p/2}$.

Proof:

Let $(\mathbb{J}_n)_{n\in\mathbb{N}}$ be a 0-sequence of partitions of Π. Then for any $n\in\mathbb{N}$

$$\sum_{J\in\mathbb{J}_n} (\Delta_J A)^2 = Y_n \cdot A_{(1,1)}, \text{ where } Y_n = \sum_{J\in\mathbb{J}_n} \Delta_J A \, 1_J.$$

Since A has no 0-jumps, $Y_n \underset{n\to\infty}{\to} 0$. Moreover, $|Y_n| \leq A_{(1,1)}$, $n\in\mathbb{N}$. Therefore, dominated convergence applies and yields the assertion. □

For martingale type processes we concentrate on the case $p\geq 2$.

Proposition 3: Let $p\geq 2$, M a p-integrable regular 2-martingale which possesses a quadratic 1-variation in $L^{p/2}$. Then: if $[M]^1 = 0$, M has quadratic variation 0 in $L^{p/2}$.
A similar result holds for 1-martingales.

Proof:

Fix a 0-sequence $(\mathbb{J}_n)_{n\in\mathbb{N}}$ of partitions of Π. Then for any $n\in\mathbb{N}$

$$(16.1) \quad \sum_{J_1\in(\mathbb{J}_n)_1} (\Delta_{J_1} M(.,1))^2 = \sum_{J\in\mathbb{J}_n} (\Delta_J M)^2 + \sum_{J\in\mathbb{J}_n} \Delta_J M \, \Delta_{J^1} M.$$

Let us estimate the last term of (16.1). Apply successively the one-parameter versions of theorem (8.2) for p/2 and of theorem (8.1) for p to obtain for $n\in\mathbb{N}$

$$(16.2) \quad \| \sum_{J\in\mathbb{J}_n} \Delta_J M \, \Delta_{J^1} M \|_{p/2}$$

$$\leq c \| [\sum_{J_2\in(\mathbb{J}_n)_2} [\sum_{J_1\in(\mathbb{J}_n)_1} \Delta_J M \, \Delta_{J^1} M]^2]^{1/2} \|_{p/2}$$

$$\leq c \| [\sum_{J\in\mathbb{J}_n} (\Delta_J M)^2 \sup_{t_2\in[0,1]} \sum_{J_1\in(\mathbb{J}_n)_1} (\Delta_{J_1} M(.,t_2))^2]^{1/2} \|_{p/2}$$

$$\leq c \| \sum_{J\in\mathbb{J}_n} (\Delta_J M)^2 \|_{p/2}^{1/2} \| \sup_{t_2\in[0,1]} \sum_{J_1\in(\mathbb{J}_n)_1} (\Delta_{J_1} M(.,t_2))^2 \|_{p/2}^{1/2}$$

$$\leq \frac{p}{p-1} c \| \sum_{J \in \mathbb{J}_n} (\Delta_J M)^2 \|_{p/2}^{1/2} \| \sum_{J_1 \in (\mathbb{J}_n)_1} (\Delta_{J_1} M_{(.,1)})^2 \|_{p/2}^{1/2},$$

where c comes from theorem (8.2). Now set

$$a_n = \| \sum_{J \in \mathbb{J}_n} (\Delta_J M)^2 \|_{p/2}, \quad b_n = \| \sum_{J_1 \in (\mathbb{J}_n)_1} (\Delta_{J_1} M_{(.,1)})^2 \|_{p/2},$$

$n \in \mathbb{N}$. Then integrating (16.1) and using (16.2) for an estimation of the right hand side gives

$$(16.3) \quad a_n \leq b_n + \frac{p}{p-1} c \, (a_n b_n)^{1/2}, \quad n \in \mathbb{N}.$$

Now assume $[M]^1 = 0$. This implies that $(b_n)_{n \in \mathbb{N}}$ is a 0-sequence. But by (16.3), $(a_n)_{n \in \mathbb{N}}$ must converge to 0, too. This means that M has quadratic variation 0 in $L^{p/2}$. □

As an immediate consequence of proposition 3, the compensators of 1- and 2- jumps of square integrable martingales on simple sets will prove not to contribute to the quadratic variation.

Proposition 4: Let $p \geq 2$, M a p-integrable continuous 2-martingale which is increasing in direction 1. Then M has quadratic variation 0 in $L^{p/2}$.

An analogous statement holds for 1-martingales which are increasing in direction 2.

Proof:

For any $t_2 \in [0,1]$, the process $M_{(.,t_2)}$ is increasing and continuous. Therefore, as is well known, $[M]^1 = 0$. Now apply proposition 3. □

Let us turn to the problem of continuity of the mapping $M \to [M]$. The following theorem shows that it is indeed continuous on $\mathcal{M}^2$ and on the space of regular $L \log^+ L$-bounded martingales.

Theorem 1: Let i=1,2.

1. Let $(M^n)_{n \in \mathbb{N}}$ be a sequence in $\mathcal{M}^2$ which converges to M. If each M^n possesses a quadratic variation in L^1, then M possesses

a quadratic variation in L^1 and for $t\in \mathbb{I}$

$$\| [M^n]_t - [M]_t \|_1 \underset{n\to\infty}{\to} 0 .$$

2. Let $(M^n)_{n\in\mathbb{N}}$ be a sequence of $L \log^+ L$-bounded regular martingales converging to M in $L \log^+ L$. If each M^n possesses a quadratic variation in L^0, then M possesses a quadratic variation in L^0 and for $t\in \mathbb{I}$

$$\| [M^n]_t - [M]_t \|_0 \underset{n\to\infty}{\to} 0.$$

Proof:

We carry out the slightly more complicated arguments for the second part. For simplicity, let $t=(1,1)$ and fix a 0-sequence $(\mathbb{J}_n)_{n\in\mathbb{N}}$ of partitions of $\mathbb{I}$. To abbreviate, set for $m\in\mathbb{N}$ and a martingale N

$$S_m(N) = \sum_{J\in\mathbb{J}_m} (\Delta_J N)^2 .$$

First note that for non-negative random variables X, Y and for $\lambda\geq 1$, $\varepsilon>0$ we have

$$(16.4) \quad \| X\, Y\|_0 \leq P(X>\lambda) + \lambda \| Y \|_0 \leq P(X>\lambda) + \lambda(\varepsilon + P(Y>\varepsilon)).$$

Apply (16.4) and the weak inequality of Burkholder (theorem (8.2)) to find a constant c independent of $(M^n)_{n\in\mathbb{N}}$ and M, such that for all $k,l,n\in\mathbb{N}$, $\lambda,\varepsilon>0$

$$(16.5) \quad \| S_k(M) - S_l(M) \|_0$$

$$\leq \| S_k(M) - S_k(M^n) \|_0 + \| S_k(M^n) - S_l(M^n) \|_0 + \| S_l(M^n) - S_l(M) \|_0$$

$$\leq 2 \sup_{k\in\mathbb{N}} \| [S_k(M+M^n)\, S_k(M-M^n)]^{1/2} \|_0 + \| S_k(M^n) - S_l(M^n) \|_0$$

$$\leq 2 \sup_{k\in\mathbb{N}} [\, P([S_k(M+M^n)]^{1/2} > \lambda) + \lambda(\varepsilon + P([S_k(M-M^n)]^{1/2} > \varepsilon))]$$

$$+ \| S_k(M^n) - S_l(M^n) \|_0$$

$$\leq c\, [\tfrac{1}{\lambda} \|(M+M^n)_{(1,1)}\|_\Psi + \lambda\varepsilon + \tfrac{\lambda}{\varepsilon} \|(M-M^n)_{(1,1)}\|_\Psi]$$

$$+ \| S_k(M^n) - S_l(M^n) \|_0 ,$$

where $\Psi(t) = t \log^+ t$, $t\geq 0$.

From (16.5) it is clear that $(S_k(M))_{k\in\mathbb{N}}$ is a Cauchy sequence in $L^0(\Omega,\mathcal{F},P)$, since $(S_k(M^n))_{k\in\mathbb{N}}$ is, for any $n\in\mathbb{N}$, and since

$$\|(M-M^n)_{(1,1)}\|_\Psi \underset{n\to\infty}{\to} 0.$$

This implies the second assertion. □

If $(M_n)_{n\in\mathbb{N}}$ is the sequence of partial sums of a series of martingales with pairwise orthogonal variation, we can give a more precise description of the quadratic variation of the limit. Let us first note that the quadratic variations of processes with orthogonal variation simply add to give the quadratic variation of the sum.

Proposition 5: Let $p\geq 0$, X and Y be processes which possess quadratic variations in L^p and have orthogonal variation. Then $X + Y$ possesses a quadratic variation in L^p and

$$[X + Y] = [X] + [Y].$$

Proof:

In case $p=0$ the assertion follows directly from the definitions. Assume $p>0$ and let $(\mathbb{J}_n)_{n\in\mathbb{N}}$ be a 0-sequence of partitions of $\mathbb{I}$. By the inequality of Cauchy-Schwarz for any $n\in\mathbb{N}$, setting

$$\xi_n = \sum_{J\in\mathbb{J}_n} \Delta_J X\, \Delta_J Y,$$

$$|\xi_n| \leq [\sum_{J\in\mathbb{J}_n} (\Delta_J X)^2 \sum_{J\in\mathbb{J}_n} (\Delta_J Y)^2]^{1/2}.$$

Since X and Y possess quadratic variations in L^p, this inequality implies that $\{|\xi_n|^p : n\in\mathbb{N}\}$ is uniformly integrable. On the other hand, X and Y have orthogonal variation. Therefore the theorem of Vitali allows to conclude that $(\xi_n)_{n\in\mathbb{N}}$ converges to 0 in L^p. This implies the assertion. □

The following special case of proposition 5 will in particular apply to compensated jumps of square integrable martingales.

Corollary: Let $p\geq 0$, X and Y be processes which possess quadratic variations in L^p. If $[Y] = 0$, then X and Y have orthogonal variation. Moreover, X + Y possesses a quadratic variation in L^p and

$$[X + Y] = [X].$$

Proof:

By an estimation like in the proof of proposition 5, $[Y] = 0$ yields that X and Y have orthogonal variation. □

Theorem 2: Let $(M^n)_{n \in \mathbb{N}}$ be a sequence in M^2 with pairwise orthogonal variation such that $\sum_{n \in \mathbb{N}} M^n = M \in M^2$. If each M^n possesses a quadratic variation in L^1, then M possesses a quadratic variation in L^1 and

$$[M] = \sum_{n \in \mathbb{N}} [M^n].$$

Moreover, if $[M^n]$ has no i-jumps for all $n \in \mathbb{N}$, then $[M]$ has no i-jumps, $i=0,1,2$.

Proof:

For $n \in \mathbb{N}$ let $N^n = \sum_{k \leq n} M^k$. By proposition 5, for any $n \in \mathbb{N}$ we know that N^n possesses a quadratic variation in L^1 and

$$[N^n] = \sum_{k \leq n} [M^k].$$

Theorem 1 implies that M possesses a quadratic variation in L^1 and

$$[M]_t = \sum_{n \in \mathbb{N}} [M^n]_t, \quad t \in \mathbb{I}.$$

But for any $n \in \mathbb{N}$ we have

$$\sup_{t \in \mathbb{I}} ([M]_t - [N^n]_t) = \sum_{k > n} [M^k]_{(1,1)}.$$

Therefore, $[M]$ inherits the continuity properties of $[M^n]$, $n \in \mathbb{N}$, and we are done. □

17. Quadratic variations of compensated components

Among the aims of this chapter is to show that square integrable regular martingales are decomposed by the pairwise orthogonal compensated jumps on simple sets of section 15 and a continuous part, and that their quadratic variation is just the sum of the quadratic variations of these components. The following two sections will provide the ingredients. In this one we will prove that the compensated jumps on simple sets possess quadratic variations and describe them.

We first consider the simplest case, the compensated 0-jumps $M(L)$ of a martingale M. The plausible result reads: the quadratic variation of such a part is just the sum of the squares of the 0-jumps on L. Here, for the first time, we see that the compensator does not change the quadratic variation of a jump part. It is somewhat more difficult to treat the 1- and 2-jumps. If, for example, we consider a compensated jump $M(L)$ for $L \in s(\mathcal{S}^1)$, we have to deal with a process which is of bounded variation in direction 1 and of generally unbounded variation in the other direction. Again, it is not hard to guess, what the quadratic variation turns out to be. It is the sum of the quadratic variations, or - what amounts to the same - the dual previsible projections $\langle \Delta^1_{s_1} M_{(.,.)} \rangle$, along the line segments L_{s_1} which constitute L, $s_1 \in [0,1]$. A similar result is obtained for $L \in s(\mathcal{S}^2)$. We finally consider continuous martingales to establish a result which has already been known for some time (see Zakai [46], Cairoli, Walsh [13], Nualart [36], [37]). Their quadratic variation is given by the dual previsible projections of their squares.

Concerning the integrability properties of the martingales considered, we will not try to attain the utmost generality in this section. The approximation theorems (16.1) and (16.2) will allow us later to pass to square integrable martingales with much less effort than would be involved in a direct proof at this stage. Our results are for martingales and simple sets which are p-integrable for any p. They are interesting in their own right since they will yield, in conjunction with the theorems of the main sections 19 and 20, approximation results for square integrable martingales.

Theorem 1: Let M be a regular martingale, $L \in S^o$ such that M and L are p-integrable for any p, and such that M(L) is compensated by a process C. Then M(L) - C possesses a quadratic variation in L^p for all p, which is given by

$$[M(L) - C] = [M(L)] = (\Delta.M)^2 \cdot \Gamma(L)\,.\,.$$

Proof:

By proposition (14.1), M(L) is p-integrable for all p, hence by theorem (4.3) and theorem (10.4) C is. Proposition (16.2) implies that C has quadratic variation 0 in L^p for all p. Therefore, by the corollary of proposition (16.5) we have to show that M(L) possesses a quadratic variation in L^p, $p \geq 0$, which is given by the above expression.

Fix a 0-sequence $(\mathbb{I}_n)_{n \in \mathbb{N}}$ of partitions of $\mathbb{I}$ and take $t=(1,1)$, for simplicity. Then for $n \in \mathbb{N}$, $p \geq 1$ by definition of M(L)

$$\Big\| \sum_{J \in \mathbb{I}_n} (\Delta_J M(L))^2 - (\Delta.M)^2 \cdot \Gamma(L)_{(1,1)} \Big\|_p$$

$$\leq 2 \, \| 1_{S_n} (\Delta.M)^2 \cdot \Gamma(L)_{(1,1)} \, \Gamma(L)_{(1,1)} \|_p ,$$

where $S_n = \bigcup_{J \in \mathbb{I}_n} \{\omega \in \Omega : |J \cap L_\omega| > 1\}$.

Since $S_n \downarrow \emptyset$, dominated convergence allows to conclude. □

If we consider 1- and 2- jumps, "partly stochastic" integrals enter the scene and the analysis becomes more involved. We prefer to start with the following auxiliary result which serves both the 1- and 2-jump case and the continuous case. It expresses the fact that the dual previsible projections of quadratic 1-variations which are continuous in the second parameter, might already be the quadratic variations of martingales. This is made plausible by the fact that the dual previsible projection of M^2 and the quadratic variation of M coincide for continuous one-parameter martingales M. To make the proof more convenient, we consider 4-integrable martingales (see Zakai [46] and Nualart [36]).

Proposition 1: Let M be a 4-integrable martingale. Assume that M is regular, possesses at most 1-jumps and fulfills

$$\Big\| \sum_{J \in \mathbb{J}_n} [(\Delta_J M)^2 - E(\Delta_J [M]^1 | \mathcal{F}^2_{s_2^J})] \Big\|_2 \underset{n\to\infty}{\to} 0$$

for any 0-sequence $(\mathbb{J}_n)_{n\in\mathbb{N}}$ of partitions of $\mathbb{I}$. Then the quadratic variation of M in L^2 exists and is given by the dual 2-previsible projection of $[M]^1$.

An analogous statement holds with respect to $[M]^2$.

Proof:

1. To show that $[M]^1$ is of class D^2, let $(\mathbb{K}_n)_{n\in\mathbb{N}}$ be a 0-sequence of partitions of $[0,1]$. Then for any 0-sequence $(\mathbb{H}_m)_{m\in\mathbb{N}}$ of partitions of $[0,1]$

$$\sum_{K\in\mathbb{K}_n} E(\Delta_K [M]^1_{(1.,)} | \mathcal{F}^2_{s^K})$$

$$= \lim_{m\to\infty} \sum_{H\times K\in\mathbb{H}_m\times\mathbb{K}_n} E(\Delta_K (\Delta_H M_{(.,.)})^2 | \mathcal{F}^2_{s^K})$$

$$= \lim_{m\to\infty} \sum_{H\times K\in\mathbb{H}_m\times\mathbb{K}_n} E((\Delta_{H\times K} M)^2 | \mathcal{F}^2_{s^K})$$

and we can apply proposition (8.4). Denote by A the dual 2-previsible projection of $[M]^1$ which exists according to proposition (16.1) and theorem (5.1).

2. Now let $(\mathbb{I}_n)_{n\in\mathbb{N}}$ be an arbitrary 0-sequence of partitions of [0,1]. For simplicity we take t=(1,1) again. We propose to prove

(17.1) $$\| \sum_{J\in\mathbb{I}_n} E(\Delta_J[M]^1_{(1,.)} | \mathcal{F}^2_{sJ}) - A_{(1,1)} \|_2 \underset{n\to\infty}{\to} 0.$$

This, together with the hypothesis, will imply the assertion. To abbreviate, set

$$\xi_J = \Delta_J[M]^1_{(1,.)}, \quad \eta_J = \Delta_J M_{(1,.)}, \quad \overline{\xi}_J = E(\xi_J | \mathcal{F}^2_{sJ}), \ J\in\mathbb{I},$$

and

$$S_n(\alpha) = \sum_{J\in\mathbb{I}_n} \alpha_J,$$

for a family α_J of random variables indexed by $J\in\mathbb{I}_n$, $n\in\mathbb{N}$. Now note that for $n,m\in\mathbb{N}$, $n\le m$,

$$\| S_n(\overline{\xi}) - S_m(\overline{\xi}) \|_2^2 = S_n(\| \overline{\xi} - \sum_{\mathbb{I}_m \ni H\subset .} \overline{\xi}_H \|_2^2)$$

$$\le 2 \, [S_n(\| \overline{\xi} \|_2^2) + S_n(\| \sum_{\mathbb{I}_m \ni H\subset .} \overline{\xi}_H \|_2^2)]$$

$$\le 2 \, [S_n(\| \eta^2 \|_2^2) + S_n(\| \sum_{\mathbb{I}_m \ni H\subset .} \eta_H^2 \|_2^2)] \quad (\overline{\xi}_J = E(\eta_J^2 | \mathcal{F}^2_{sJ}), \text{ theorem (4.3)})$$

$$\le c_1 \, S_n(\| \eta \|_4^4) \quad \text{(theorem (8.2))}$$

$$\le c_1 \, \| \sup_{J\in\mathbb{I}_n} |\eta_J| \|_4^2 \, \| S_n(\eta^2) \|_2 \quad \text{(Cauchy-Schwarz)}$$

$$\le c_2 \, \| \sup_{J\in\mathbb{I}_n} |\eta_J| \|_4^2 \, \| M_{(1,1)} \|_4 \quad \text{(theorem (8.2))},$$

where the constants c_1, c_2 result from the application of theorem (8.2). Now M has at most 1-jumps. Therefore, dominated convergence on the basis of theorem (8.1) implies that $(S_n(\overline{\xi}))_{n\in\mathbb{N}}$ is a Cauchy sequence in $L^2(\Omega,\mathcal{F},P)$. Hence, (17.1) is a consequence of theorem (5.2). This completes the proof. □

We return to the computation of the quadratic variations of compensated jumps on simple sets.

<u>Theorem 2</u>: Let M be a regular martingale without 0-jumps,

$L \in s(\mathcal{S}^1)$, such that M, L are p-integrable for any p, and such that M(L) is compensated by a process C. Then M(L) - C possesses a quadratic variation in L^p for all p, which is given by

$$[M(L) - C]_t = [M(L)]_t =$$

$$= \int_{[0,t]} [<\Delta^1_{s_1} M_{(.,.)}>_{t_2} - <\Delta^1_{s_1} M_{(.,.)}>_{s_2}]\, d\Gamma(\partial L)_s, \quad t \in \mathbb{I},$$

if $L \in \mathcal{S}^1$, and by a difference of two expressions like this in the general case. A corresponding statement holds with respect to the second parameter.

Proof:

1. We first assume that $L \in \mathcal{S}^1$. Proposition (14.2) shows that M(L) is p-integrable, theorem (5.3) does the same for C, for any $p \geq 0$.

To compute $[M(L)]^1$, fix a 0-sequence $(\mathbb{K}_n)_{n \in \mathbb{N}}$ of partitions of [0,1]. Let t=(1,1) again. Then for any $n \in \mathbb{N}$, $p \geq 1$

$$\| \sum_{K \in \mathbb{K}_n} (\Delta_K M(L)_{(.,1)})^2 - \int_{\mathbb{I}} (\Delta_{]s_2,1]} \Delta^1_{s_1} M_{(.,.)})^2\, d\Gamma(\partial L)_s \|_p$$

$$\leq 2\, \| 1_{T_n}\, 16 \sup_{t \in \mathbb{I}} |M_t|^2\, \Gamma(\partial L)^2_{(1,1)} \|_p ,$$

where $T_n = \bigcup_{K \in \mathbb{K}_n} \{\omega \in \Omega: |J \times [0,1] \cap \partial L_\omega| > 1\}$. Since $T_n \downarrow \emptyset$, dominated convergence gives that M(L) has a quadratic 1-variation in L^p for all p, which is given by the process

$$\int_{[0,t]} (\Delta_{]s_2,t_2]} \Delta^1_{s_1} M_{(.,.)})^2\, d\Gamma(\partial L)_s , \quad t \in \mathbb{I} .$$

Since C is continuous and of bounded variation in direction 1, this is also the quadratic 1-variation of M(L) - C.

2. To verify the hypothesis of proposition 1, we prove

$$(17.2) \quad \| \sum_{J \in \mathbb{J}_n} [(\Delta_J M(L))^2 - E((\Delta_J M(L))^2 | \mathcal{F}^2_{s_2^J})] \|_2 \underset{n \to \infty}{\to} 0 ,$$

$$(17.3) \quad \| \sum_{J \in \mathbb{J}_n} [E((\Delta_J M(L))^2 | \mathcal{F}^2_{s_2^J}) - E(\Delta_J [M(L)]^1 | \mathcal{F}^2_{s_2^J})] \|_2 \underset{n \to \infty}{\to} 0 ,$$

for any 0-sequence $(\mathbb{J}_n)_{n \in \mathbb{N}}$ of partitions of $\mathbb{I}$. (17.2) and

(17.3) will imply the hypothesis of proposition 1, since C has quadratic variation 0 by proposition (16.4). Note first that for $J \in \mathcal{J}$, setting

$$\xi_J = \int_J \Delta_{]s_2,t_2]}\Delta^1_{s_1} M_{(.,.)} \, d\Gamma(\partial L)_s ,$$

$$\eta_J = \int_{J^1} \Delta_{J_2}\Delta^1_{s_1} M_{(.,.)} \, d\Gamma(\partial L)_s ,$$

we have

$$\Delta_J M(L) = \xi_J + \eta_J .$$

To abbreviate, we set for any family α_J of random variables, indexed by $J \in \mathbb{I}_n$, $n \in \mathbb{N}$,

$$\bar{\alpha}_J = E(\alpha_J | \mathcal{F}^2_{s^J_2}), \quad S^n(\alpha) = \sum_{J \in \mathbb{I}_n} \alpha_J , \quad S^n_1(\alpha)_{J_2} = \sum_{J_1 \in (\mathbb{I}_n)_1} \alpha_J ,$$

$S^n_2(\alpha)_{J_1}$ analogously.

To treat the first term in the above decomposition of $\Delta.M(L)$, observe that

$$S^n(\xi^2) = Y_n \cdot \Gamma(\partial L)_{(1,1)} ,$$

where $Y_n = \sum_{J \in \mathbb{I}_n} \xi_J \Delta_{]\cdot,t^J_2]} \Delta^1_{\cdot} M_{(.,.)} 1_J$, $n \in \mathbb{N}$. Since M has no 0-jumps, $Y_n \underset{n\to\infty}{\to} 0$, dominated by $16 \sup_{t \in \Pi} |M_t|^2 \, \Gamma(\partial L)_{(1,1)}$. Therefore, dominated convergence on the basis of theorem (8.1) gives

$$\| S^n(\xi^2) \|_2 \underset{n\to\infty}{\to} 0,$$

and (17.2) will follow from

$$(17.4) \quad \| S^n(\eta^2 - \overline{\eta^2}) \|_2 \underset{n\to\infty}{\to} 0.$$

We note that $\{S^n_1(\eta^2 - \overline{\eta^2})_{J_2} : J_2 \in (\mathbb{I}_n)_2\}$ is a family of martingale differences for any $n \in \mathbb{N}$. Hence

$$(17.5) \quad \| S^n(\eta^2 - \overline{\eta^2}) \|^2_2 = S^n_2(\| S^n_1(\eta^2 - \overline{\eta^2}) \|^2_2)$$

$$\leq 2[S^n_2(\| S^n_1(\eta^2) \|^2_2) + S^n_2(\| S^n_1(\overline{\eta^2}) \|^2_2)]$$

$$\leq 4 \, S^n_2(\| S^n_1(\eta^2) \|^2_2) \qquad \text{(Jensen's inequality)}$$

$$\leq 4 \, \| \sup_{J_2 \in (\mathbb{I}_n)_2} S^n_1(\eta^2)_{J_2} \|_2 \, \| S^n(\eta^2) \|_2 , \quad n \in \mathbb{N}.$$

To estimate the right hand side of (17.5), observe that since M has no 0-jumps and ∂L finite $\mathbb{I}$-sections,

$$\sup_{J_2 \in (\mathbb{J}_n)_2} S_1^n(\eta^2)_{J_2} \underset{n\to\infty}{\to} 0,$$

dominated by 16 $\sup_{t\in\mathbb{I}} |M_t|^2 \; \Gamma(\partial L)^2_{(1,1)}$. Therefore, the first term converges to 0 as $n\to\infty$, whereas the second remains bounded in virtue of theorem (8.2) applied to $M(L) - C$ and proposition (16.4), applied to C. This proves (17.4) and consequently (17.2). To show (17.3), we first remark that we can discard a "small contribution" like the ξ's above. Then, setting

$$\rho_J = \int_{J^1} (\Delta_{J_2} \Delta^1_{s_1} M_{(.,.)})^2 \; d\Gamma(\partial L)_s, \; J\in\mathcal{J},$$

we have

(17.6)

$$E(\int_{J^1} [(\Delta_{]s_2,t_2^J]} \Delta^1_{s_1} M_{(.,.)})^2 - (\Delta_{]s_2,s_2^J]} \Delta^1_{s_1} M_{(.,.)})^2] \; d\Gamma(\partial L)_s | \mathcal{F}^2_{s_2^J})$$
$$= \bar{\rho}_J = E(\int_{J^1} \Delta_{]s_2^J,t_2^J]} \langle \Delta^1_{s_1} M_{(.,.)}\rangle \; d\Gamma(\partial L)_s | \mathcal{F}^2_{s_2^J})$$

and

$$\| S^n(\overline{\eta^2} - \bar{\rho}) \|_2 \le 2 \| S^n(|\eta^2 - \rho|) \|_2 \qquad \text{(theorem (4.3))}$$
$$\le 2 \; \| 1_{T_n} \; 4 \sup_{s_1\in[0,1]} \sum_{J_2\in(\mathbb{J}_n)_2} (\Delta_{J_2} M_{(s_1,.)})^2 \; \Gamma(\partial L)^2_{(1,1)} \|_2 ,$$

where T_n is defined in part 1, $n\in\mathbb{N}$. This, however, converges to 0 by dominated convergence on the basis of theorems (8.1) and (8.2) this time. An appeal to (17.6) yields (17.3). This finishes the verification of the hypothesis of proposition 1, which thus can be applied. For the p-integrability of the quadratic variation, theorem (8.2) once again applies. It is easy to identify the dual 2-previsible projection of $[M(L) - C]^1$ with the process of the assertion: look at (17.6) and theorem (5.2).

3. Finally suppose $\mathcal{S}^1 \ni K \subset L \in \mathcal{S}^1$. Then an argument which primarily uses the 2-martingale property of $M(K)$, $M(L)$ and no new idea

otherwise, gives $[M(L\setminus K)] = [M(L)] - [M(K)]$. This completes the proof. □

Another application of proposition 1 will now give us the existence of the quadratic variation of continuous martingales.

Theorem 3: Let M be a continuous martingale which is p-integrable for any p. Then M possesses a quadratic variation in L^p for all p, which is given by $\langle M\rangle$.

Proof:

From theorem (5.2) and the one-parameter result which was implicitly used in the preceding proof (see Métivier [31], p. 122) we know that $[M]^i = \langle M\rangle^i$, $i=1,2$. Therefore, theorem (11.2) implies the assertion, once we have established the hypothesis of proposition 1. Let $(\mathbb{I}_n)_{n\in\mathbb{N}}$ be a 0-sequence of partitions of $\mathbb{I}$. To abbreviate, for $J\in\mathbb{I}_n$, $n\in\mathbb{N}$, set

$$\xi_J = (\Delta_J M)^2 , \quad \eta_J = \Delta_J [M]^1 ,$$

and for a family α_J of random variables, indexed by $J\in\mathbb{I}_n$, $S^n(\alpha)$, $S_1^n(\alpha)_{J_2}$, $S_2^n(\alpha)_{J_1}$ as in the preceding proof,

$$\overline{\alpha}_J = E(\alpha_J | F_{s^J}), \quad \overline{\alpha}_J^i = E(\alpha_J | F^i_{s_i^J}), \quad i=1,2.$$

We will show

$$(17.7) \quad \| S^n(\xi - \overline{\xi}^2) \|_2 \underset{n\to\infty}{\to} 0,$$

$$(17.8) \quad \| S^n(\overline{\xi}^2 - \overline{\xi}) \|_2 \underset{n\to\infty}{\to} 0,$$

$$(17.9) \quad \| S^n(\overline{\eta}^2 - \overline{\eta}) \|_2 \underset{n\to\infty}{\to} 0.$$

These statements together with $\overline{\xi}_J = \overline{\eta}_J$, $J\in\mathbb{I}_n$, $n\in\mathbb{N}$, which follows from (9.2), will establish the hypothesis of proposition 1. Now the proof of (17.9) is almost identical to the proof of (17.8) and the discrete version of theorem (4.3) gives

$$\| S^n(\overline{\xi}^2 - \overline{\xi}) \|_2$$
$$\le 2 \| S^n(\xi - \overline{\xi}^1) \|_2 , \quad n\in\mathbb{N}.$$

We may therefore confine ourselves to the proof of (17.7). But the first arguments are exactly the same as for (17.5). Thus, skipping the first steps, we have for $n \in \mathbb{N}$

$$\| S^n(\xi - \bar{\xi}^2) \|_2^2$$
$$\leq 4 \quad S_2^n(\| S_1^n(\xi) \|_2^2)$$
$$\leq c \sum_{J_2 \in (\mathbb{I}_n)_2} \| \Delta_{J_2} M_{(1,.)} \|_4^4 \qquad \text{(theorem (8.2))}.$$

Now we can conclude like at the end of the proof of proposition 1, replacing the η_J there by $\Delta_{J_2} M_{(1,.)}$. □

18. Orthogonality of compensated components

We will now establish that compensated jump components belonging to pairwise disjoint simple sets of the same type or to simple sets of different types, including continuous martingales, have pairwise orthogonal variation. For two simple sets of different types or one compensated jump and a continuous martingale this is a consequence of the essential difference of the various "continuity degrees".

Since we already know that square integrable martingales with orthogonal variation are orthogonal in M^2, these results will provide the last ingredients for both the structure theorem for square integrable martingales and their quadratic variations. Like in the preceding section, they will be stated for p-integrable martingales for all p only.

At first, we compare two different 0-jump components.

Proposition 1: Let M, N be regular martingales, K, $L \in S^o$ disjoint sets such that M, N, K, L are p-integrable for any p, and such that $M(K)$ resp. $N(L)$ is compensated by C resp. D. Then $M(K) - C$ and $N(L) - D$ have orthogonal variation.

Proof:

According to proposition (16.2) and the corollary of proposition (16.5), all we have to verify is that $M(K)$ and $N(L)$ have orthogonal variation. To do this, let an arbitrary 0-sequence $(\mathbb{I}_n)_{n\in\mathbb{N}}$ of partitions of $\mathbb{I}$ be given. For $n\in\mathbb{N}$, set

$$H_n = \{J\in\mathbb{I}_n : J\cap K \neq \emptyset\}, \quad I_n = \{J\in\mathbb{I}_n : J\cap L \neq \emptyset\}.$$

Then the sequence of random variables $(|H_n \cap I_n|)_{n\in\mathbb{N}}$ converges to 0, since K and L are disjoint. Consequently,

$$\lim_{n\to\infty} \sum_{J\in\mathbb{I}_n} |\Delta_J M(K)\ \Delta_J N(L)| = \lim_{n\to\infty} \sum_{J\in H_n\cap I_n} |\Delta_J M(K)\ \Delta_J N(L)| = 0,$$

at least in $L^0(\Omega,\mathcal{F},P)$. The assertion follows. □

Next, we compare a 0-jump component to a jump component of a different type or a continuous martingale.

Proposition 2: Let M, N be regular martingales, $L\in S^0$ such that M, N, L are p-integrable for all p, and such that $M(L)$ is compensated by a process C. Moreover, assume N has no 0-jumps. Then $M(L) - C$ and N have orthogonal variation.

Proof:

We have to show that $M(L)$ and N have orthogonal variation. Let $(\mathbb{I}_n)_{n\in\mathbb{N}}$ be an arbitrary 0-sequence of partitions of $\mathbb{I}$. For any $n\in\mathbb{N}$, by the definition of $M(L)$

$$\sum_{J\in\mathbb{I}_n} |\Delta_J M(L)\ \Delta_J N| \leq Y_n \cdot \Gamma(L)_{(1,1)},$$

where $Y_n = \sum_{J\in\mathbb{I}_n} |\Delta_J N|\,|\Delta.M|\ 1_J$. Since N has no 0-jumps, obviously $Y_n \underset{n\to\infty}{\to} 0$, dominated by $16 \sup_{t\in\mathbb{I}} |M_t| \sup_{t\in\mathbb{I}} |N_t|$. Therefore, dominated convergence applies and gives the desired result. □

Comparing two 1-jump components with disjoint jump sets or a 1- and a 2-jump component involves a simple idea like for proposition 1.

Proposition 3: Let M, N be regular martingales without 0-jumps, K, $L \in s(S^1) \cup s(S^2)$ such that M, N, K, L are p-integrable for any p, and such that M(K) resp. N(L) is compensated by a process C resp. D. Moreover, assume that K and L are disjoint if they are of the same kind. Then M(K) - C and N(L) - D have orthogonal variation.

Proof:
According to proposition (16.4) and the corollary of proposition (16.5) we have to prove that M(K) and N(L) have orthogonal variation. Fix a 0-sequence $(\mathbb{J}_n)_{n \in \mathbb{N}}$ of partitions of $\mathbb{I}$ and define

$$H_n = \{J \in \mathbb{J}_n : \; J \cap K \neq \emptyset\} , \quad I_n = \{J \in \mathbb{J}_n : \; J \cap L \neq \emptyset\} .$$

Then the sequence of random variables $(|H_n \cap I_n|)_{n \in \mathbb{N}}$ converges to the random finite number of points in the intersection of the closures of K and L. Therefore, since M and N have no 0-jumps,

$$\lim_{n \to \infty} \sum_{J \in \mathbb{J}_n} |\Delta_J M(K) \; \Delta_J N(L)| = \lim_{n \to \infty} \sum_{J \in H_n \cap I_n} |\Delta_J M(K) \; \Delta_J N(L)| = 0 ,$$

at least in $L^0(\Omega, \mathcal{F}, P)$. The proof is complete. □

We finally have to compare 1- or 2-jump components to continuous martingales.

Proposition 4: Let M, N be regular martingales without 0-jumps, $L \in s(S^1) \cup s(S^2)$ such that M, N, L are p-integrable for any p, and such that M(L) is compensated by a process C. Moreover, assume N is continuous. Then M(L) - C and N have orthogonal variation.

Proof:
It is enough to show that M(L) and N have orthogonal variation. We may assume $L \in S^1$. Let $(\mathbb{J}_n)_{n \in \mathbb{N}}$ be a 0-sequence of partitions of $\mathbb{I}$. If for $J \in \mathcal{J}$ we set

$$\xi_J = \int_J \Delta_{]s_2,t_2^J]} \Delta^1_{s_1} M(.,.)\; d\Gamma(\partial L)_s \quad ,$$

$$\eta_J = \int_{J^1} \Delta_{J_2} \Delta^1_{s_1} M(.,.)\; d\Gamma(\partial L)_s \quad ,$$

we have

$$\Delta_J M(L) = \xi_J + \eta_J .$$

Now for any $n \in \mathbb{N}$

$$\sum_{J \in \mathbb{J}_n} |\Delta_J N\, \xi_J| \le Y_n \cdot \Gamma(\partial L)_{(1,1)} \quad ,$$

where $Y_n = \sum_{J \in \mathbb{J}_n} |\Delta_J N| \, |\Delta_{].,t_2^J]} \Delta^1_. M(.,.)| \, 1_J$. But $Y_n \underset{n\to\infty}{\to} 0$, since N is continuous, dominated by $16 \sup_{t\in \Pi} |M_t| \sup_{t \in \Pi} |N_t|$. Dominated convergence applies and yields

$$\sum_{J \in \mathbb{J}_n} |\Delta_J N\, \xi_J| \underset{n\to\infty}{\to} 0 \quad \text{in } L^0(\Omega, \mathcal{F}, P).$$

Let us now consider the η-terms in the above decomposition of $\Delta . M(L)$. For $n \in \mathbb{N}$ we have

$$\| \sum_{J \in \mathbb{J}_n} |\Delta_J N\, \eta_J| \|_1$$

$$= \| \sum_{J \in \mathbb{J}_n} | \int_{J^1} \Delta_J N\, \Delta_{J_2} \Delta^1_{s_1} M(.,.)\; d\Gamma(\partial L)_s | \|_1$$

$$\le \| \sum_{J \in \mathbb{J}_n} \int_{J^1} (\Delta_J N)^2\, d\Gamma(\partial L)_s \|_1^{1/2}$$

$$\cdot \| \sum_{J \in \mathbb{J}_n} \int_{J^1} (\Delta_{J_2} \Delta^1_{s_1} M(.,.))^2\, d\Gamma(\partial L)_s \|_1^{1/2}$$

$$\le 4 \; \| Z_n \cdot \Gamma(\partial L)_{(1,1)} \|_1^{1/2} \; \| \sup_{t \in \Pi} |M_t|^2 \; \Gamma(\partial L)_{(1,1)} \|_1^{1/2} \quad ,$$

where

$$Z_n = \sum_{K \in \mathbb{J}_n} (\Delta_{K_1 \times]t_2^K, 1]} N)^2 \, 1_K$$

and the last inequality results from the martingale properties of M and N and a simple change of summation replacing J_2 by K_2 such that $J_1 \times K_2 \subset J^1$. Now $Z_n \cdot \Gamma(\partial L)_{(1,1)} \underset{n\to\infty}{\to} 0$, since N is continuous and ∂L has finite Π-sections. This convergence being dominated by 16 $\sup_{t\in\Pi} |N_t|^2 \, \Gamma(\partial L)_{(1,1)}$, theorem (8.1) allows us to conclude that

$$\sum_{J\in\mathbb{J}_n} |\Delta_J N \, \eta_J| \underset{n\to\infty}{\to} 0 \quad \text{in } L^1(\Omega,\mathcal{F},P).$$

This completes the proof. □

19. The structure of square integrable martingales

Let us briefly recollect some important elements of our theory in chapters III and IV. Section 14 taught us that square integrable regular martingales have their jumps on countably many sets in $s(S^i)$, i=0,1,2, which are pairwise disjoint and of one of the following types: previsible or 1-previsible, 2-inaccessible or 2-previsible, 1-inaccessible or totally inaccessible for the 0-jumps, previsible or inaccessible for the i-jumps, i=1,2. In section 15 we learnt how the jumps of the martingales on sets of these types can be compensated by processes of essentially better continuity properties. The resulting compensated jump components and continuous martingales are pairwise orthogonal, at least for martingales which are integrable for any power, according to section 18. On the basis of all these results we will see in this section, how we can define "i-jump parts" M^i, i=0,1,2, and a continuous part M^c for square integrable martingales M, such that M decomposes into the pairwise orthogonal components M^i and M^c, i=0,1,2.
We proceed in three steps. At first, we consider a bounded martingale M and split it into a part M^o which receives all its 0-jumps and a part N which is free of them. We define M^o to be

the sum of all compensated jumps on pairwise disjoint 0-simple sets which are p-integrable for all p. Since the compensators have no more 0-jumps, this leaves $N = M - M^o$ with at most 1- or 2-jumps, but still p-integrable for all p - although not necessarily bounded - and therefore orthogonal to M^o. In the second step we consider N and define M^1 resp. M^2 to be the sum of all compensated jumps on pairwise disjoint sets in $s(\mathcal{S}^1)$ resp. $s(\mathcal{S}^2)$, which are p-integrable for all p. Then M^1 and M^2 are orthogonal and, since the compensators are continuous, $M^c = N - M^1 - M^2$ is continuous and p-integrable for all p. To sum up after two steps: a bounded martingale M can be represented by jump parts M^o, M^1, M^2 and a continuous part M^c which are p-integrable for all p; the i-jump part M^i is given by the orthogonal sum of all compensated i-jumps of M on a sequence of pairwise disjoint sets in $s(\mathcal{S}^i)$, i=0,1,2; the representation is orthogonal and unique.

In the third step, we only have to extend this result to all martingales of the Hilbert space $\mathcal{M}^2$. Since the linear space of bounded regular martingales is dense in $\mathcal{M}^2$ and the mappings $M \to M^i$, i=0,1,2, defined on this space, have the properties of projectors, this is nothing but a rather routine extension procedure.

Proposition 1: Let M be a bounded regular martingale. Then there exist unique regular martingales M^o and N which are p-integrable for any $p \geq 0$ such that

i) for any enumeration $(L_n)_{n \in \mathbb{N}}$ of 0-simple sets according to theorem (14.1), i) - iv),

$$M^o = \sum_{n \in \mathbb{N}} (M(L_n) - C_n) ,$$

where C_n is the compensator of $M(L_n)$, $n \in \mathbb{N}$,

ii) N has no 0-jumps,

iii) M^o and N are orthogonal,

iv) $M = M^o + N$.

Proof:

Let $(L_n)_{n\in\mathbb{N}}$ be an enumeration of the sequences of 0-simple sets according to theorem (14.1), i)-iv), C_n the compensator of $M(L_n)$ according to theorem (15.1) or (15.2), $n\in\mathbb{N}$. For $k\in\mathbb{N}$ let

$$Q^k = \sum_{n\leq k} (M(L_n) - C_n).$$

Then for $k\in\mathbb{N}$, $p>1$

$$(19.1)\quad \| Q^k_{(1,1)} \|_p$$

$$\leq c_1 \| \sum_{n\leq k} [M(L_n)]_{(1,1)} \|_{p/2}^{1/2} \qquad \text{(theorems (8.2), (17.1), proposition (18.1))}$$

$$\leq c_1 \| \liminf_{m\to\infty} \sum_{J\in\mathcal{J}_m} (\Delta_J M)^2 \|_{p/2}^{1/2}$$

$$\leq c_2 \| M_{(1,1)} \|_p \qquad \text{(Fatou's lemma, theorem (8.2))},$$

where c_1, c_2 are combinations of constants resulting from theorem (8.2), and $(\mathcal{J}_m)_{m\in\mathbb{N}}$ is an arbitrary sequence of partitions of Π. Since the components of Q^k are pairwise orthogonal by proposition (8.3) and theorem (17.1), the inequality (19.1) in particular implies that $(Q^k_{(1,1)})_{k\in\mathbb{N}}$ is a Cauchy sequence in $L^2(\Omega,\mathcal{F},P)$. Theorem (9.2) even says that $(Q^k)_{k\in\mathbb{N}}$ is a Cauchy sequence in $\mathcal{M}^2$. Hence

$$M^o = \sum_{n\in\mathbb{N}} (M(L_n) - C_n)$$

is a well defined element of $\mathcal{M}^2$, which, in virtue of (19.1), is p-integrable for any $p\geq 0$, as is $N = M - M^o$. Moreover, since M has no 0-jumps outside $\bigcup_{n\in\mathbb{N}} L_n$ and since the compensators are free of 0-jumps, N has no 0-jumps. Now proposition (18.2) implies that N and M^o have orthogonal variation, hence are orthogonal. This has uniqueness as another consequence. The proof is complete. □

We next extract the 1- and 2-jump parts from the remainder N.

Proposition 2: Let M be a regular martingale without 0-jumps which is p-integrable for any p. Then there exist unique regular martingales M^1, M^2 and M^c which are p-integrable for any p such that

i) for any enumeration $(L_n^i)_{n\in\mathbb{N}}$ of sets in $s(S^i)$ according to theorem (14.2), i) - iv),

$$M^i = \sum_{n\in\mathbb{N}} (M(L_n^i) - C_n^i),$$

where C_n^i is the compensator of $M(L_n^i)$, $n\in\mathbb{N}$, i=1,2,

ii) M^c is continuous, M^i has no 0-jumps and no $\bar{i}$-jumps, i=1,2,

iii) M^1, M^2, M^c are pairwise orthogonal,

iv) $M = M^1 + M^2 + M^c$.

Proof:

To extract M^1, we omit the upper index 1 in i) and suppose $(L_n)_{n\in\mathbb{N}}$ to be an enumeration of the sequences of sets according to theorem (14.2), i), ii), C_n the compensator of $M(L_n)$ according to theorem (15.3), $n\in\mathbb{N}$. If for $k\in\mathbb{N}$ we let again

$$Q^k = \sum_{n\leq k} (M(L_n) - C_n),$$

we can show in a similar way as in the proof of proposition 1, using theorem (17.2) instead of (17.1), and theorems (8.1) and (8.2) that for $k\in\mathbb{N}$, $p>1$

$$\| Q^k_{(1,1)} \|_p \leq c \; \| M_{(1,1)} \|_p ,$$

where c is a constant depending on the constants of theorems (8.1) and (8.2). We find unique regular martingales M^1 and N which are p-integrable for all $p\geq 0$ such that

i') for any enumeration $(L_n)_{n\in\mathbb{N}}$ of the sequences of sets in $s(S^1)$ according to theorem (14.2), i), ii)

$$M^1 = \sum_{n\in\mathbb{N}} (M(L_n) - C_n),$$

where C_n is the compensator of $M(L_n)$, $n\in\mathbb{N}$, ii') N has no 0-jumps and no 1-jumps, M^1 no 0-jumps and no 2-jumps,

iii') M^1 and N are orthogonal,

iv') $M = M^1 + N$.

It is clear how to apply the procedure just executed once more to N to extract a 2-jump part M^2. The remainder $N - M^2$ has no 0-jumps and no 1- or 2-jumps, hence is continuous. The orthogonality of M^1 and M^2 follows from proposition (18.3), the orthogonality of M^c and M^1 or M^2 from proposition (18.4), always in conjunction with proposition (8.3). □

Before we extend the above decomposition results to square integrable martingales, let us combine them to obtain a complete representation of bounded martingales.

<u>Proposition 3</u>: Let M be a bounded regular martingale. Then there exist unique regular martingales M^o, M^1, M^2, M^c which are p-integrable for any $p \geq 0$ such that

i) for any enumeration $(L_n^i)_{n \in \mathbb{N}}$ of sets in $s(S^i)$ according to theorem (14.1), i)-iv) resp. theorem (14.2), i)-iv),

$$M^o = \sum_{n \in \mathbb{N}} (M(L_n^o) - C_n^o) \ , \quad M^i = \sum_{n \in \mathbb{N}} ((M-M^o)(L_n^i) - C_n^i), \ i=1,2,$$

where C_n^i is the compensator of $M(L_n^o)$ resp. $(M-M^o)(L_n^i)$, $i=1,2$, $n \in \mathbb{N}$,

ii) M^c is continuous, M^i has no 0-jumps and no $\bar{i}$-jumps, $i=1,2$,

iii) M^o, M^1, M^2, M^c are pairwise orthogonal,

iv) $M = M^o + M^1 + M^2 + M^c$.

<u>Proof</u>:

Combine propositions 1 and 2. For orthogonality of M^o and the remaining three components consult i), proposition (18.2) and proposition (8.3). □

We can state our results in a more sophisticated form, which, however, exploits the important fact that M^2 is a Hilbert space and therefore makes the extension of proposition 3 to M^2 easier.

Remark: Denote by M^∞ the linear subspace of M^2 consisting of the p-integrable martingales for all p. Let $L \in S^o \cup s(S^1) \cup s(S^2)$ be a set of one of the types of theorems (14.1) or (14.2) which is p-integrable for all $p \geq 0$. Then the linear mappings

$$S^i : M^\infty \to M^2, \quad M \to M^i, \quad i=0,1,2,c,$$

$$S(L) : M^\infty \to M^2, \quad M \to \begin{cases} M(L) - C & , \text{ if } L \in S^o, \\ (M-M^o)(L) - C, & \text{if } L \notin S^o, \end{cases}$$

where C is the compensator of M(L) according to theorems (15.1)-(15.3), possess the following properties. They are idempotent by the continuity properties of the compensators and symmetric by the orthogonality results of section 18 and proposition (8.3). Moreover, their norms are bounded by 1. Therefore their unique extensions to M^2 which exist since M^∞ is dense in M^2 are idempotent and self-adjoint. But this means that they are orthogonal projectors (see Weidmann [44], pp. 74, 83).

Definition 1: Let $L \in S^o \cup s(S^1) \cup s(S^2)$ be a set of one of the types of theorems (14.1) or (14.2) which is p-integrable for any p. The unique extensions T^i, $T(L)$ of S^i, $S(L)$, $i=0,1,2,c$, according to the preceding remark are called successively "i-jump part", i=0,1,2, "continuous part", "compensated jump on L".

Now we can state the main result of this section.

Theorem 1: Let $M \in M^2$. Then there exist unique martingales $M^o, M^1, M^2, M^c \in M^2$ such that

i) for any enumeration $(L_n^i)_{n \in \mathbb{N}}$ of sets in $s(S^i)$ according to theorem (14.1), i)-iv) resp. theorem (14.2), i)-iv),

$$M^i = T^i M = \sum_{n \in \mathbb{N}} T(L_n^i) M, \quad i=0,1,2,$$

ii) $M^c = T^c M$ is continuous, M^i has no 0-jumps and no $\bar{i}$-jumps, i=1,2,

iii) M^o, M^1, M^2, M^c are pairwise orthogonal,

iv) $M = M^o + M^1 + M^2 + M^c$.

Proof:

By definition and proposition 3, i)-iv) hold on the space of bounded regular martingales which is dense in $\mathcal{M}^2$. Moreover,

$T(K)\ T(L) = 0$, if K,L are disjoint,

$T^i\ T^j = 0$, if $i,j \in \{0,1,2,c\}$, $i \neq j$.

Also, the subspaces of $\mathcal{M}^2$ consisting of the continuous martingales or of the martingales which have no 0-jumps and no $\bar{1}$-jumps, are closed. Therefore, i)-iv) extend to $\mathcal{M}^2$. □

Definition 2: Let $M \in \mathcal{M}^2$. The unique martingales M^i, i=0,1,2,c, according to theorem 1, are called successively "i-part of M", i=0,1,2, "continuous part of M".

20. The quadratic variation of square and Llog$^+$L-integrable martingales

The fact that in section 18 we established not only orthogonality but orthogonality of the variations of compensated jump parts, will now pay off. With its help we can complete the decomposition theorems for square integrable martingales of section 19 with analogous decomposition theorems for their quadratic variations. Since in section 17 a description of the quadratic variations of the jump components was given whose continuity properties are obvious, we also obtain information on the continuity properties of [M] for square integrable martingales M.

For this sake, we first have to complete our knowledge of

the continuity of the components by a simple result concerning M^c. We proceed to state the analogue of proposition (19.3) for quadratic variations of bounded martingales. The approximation theorem (16.1) will then allow us to prove the first main theorem of this section, the decomposition theorem for the quadratic variations of square integrable martingales. For $L\log^+L$-bounded martingales we at least obtain the existence of quadratic variation, as follows from the second part of theorem (16.1). But, in lack of a structure theorem for this larger class of martingales, we are unable to give a detailed description of their quadratic variations.

Proposition 1: Let M be a continuous square integrable martingale. Then $\langle M\rangle$ is continuous.

Proof:

Assume that in the decomposition of $\langle M\rangle$ according to theorem (12.1) one of the jump parts of $\langle M\rangle$ is non-zero. Pick a set $L \in S^o \cup s(S^1) \cup s(S^2)$ such that $\langle M\rangle(L) \neq 0$. We know that $L \in \mathcal{P}$. But, due to the continuity of M, M(L), as a stochastic integral process of M, is continuous, hence vanishes. Thus we have a contradiction with the isometry statement of proposition (9.2). Remember $m_{M^2} = m_{\langle M\rangle}$. □

For bounded martingales, we can describe the quadratic variation by a series of processes as computed in section 17.

Proposition 2: Let M be a bounded regular martingale. Then M, M^o, M^1, M^2, M^c possess quadratic variations in L^p for any $p\geq 0$, which satisfy

i) for any enumeration $(L_n^i)_{n\in\mathbb{N}}$ of sets in $s(S^i)$ according to theorem (14.1), i)-iv) resp. theorem (14.2), i)-iv)

$$[M^o] = \sum_{n\in\mathbb{N}} [M(L_n^o)] \ , \ [M^i] = \sum_{n\in\mathbb{N}} [(M-M^o)(L_n^i)] \ , \ i=1,2,$$

ii) $[M^c] = \langle M^c \rangle$ is continuous, $[M^i]$ has no 0-jumps and no $\bar{1}$-jumps, $i=1,2$,

iii) M^o, M^1, M^2, M^c have pairwise orthogonal variation,

iv) $[M] = [M^o] + [M^1] + [M^2] + [M^c]$.

Proof:

Let $(L_n)_{n\in\mathbb{N}}$ be an enumeration of the 0-simple sets according to theorem (14.1), i)-iv), C_n the compensator of $M(L_n)$, $n\in\mathbb{N}$. For $k\in\mathbb{N}$ let

$$Q^k = \sum_{n\leq k} (M(L_n) - C_n).$$

Apply theorem (17.1), proposition (18.1) and proposition (16.5) to see that Q^k possesses a quadratic variation in L^p for any $p\geq 0$ which is given by

$$[Q^k] = \sum_{n\leq k} [M(L_n) - C_n] = \sum_{n\leq k} [M(L_n)].$$

Now $(Q^k)_{k\in\mathbb{N}}$ converges to M^o in L^p for any $p\geq 0$. By theorem (16.2), M^o possesses a quadratic variation in L^1, hence in L^p for all $p\geq 0$ by theorem (8.2), which is given by

$$[M^o] = \sum_{n\in\mathbb{N}} [M(L_n)].$$

In a similar way, using theorem (17.2), proposition (18.3) and proposition (16.5), we can show that M^1 and M^2 possess quadratic variations in L^p for all $p\geq 0$ such that i) is satisfied. An appeal to theorem (17.3), proposition 1 and theorem (16.2) also yields ii). Since orthogonality of variations is preserved by convergence in $\mathcal{M}^2$ in virtue of theorem (8.2), the propositions (18.2), (18.3) and (18.4) imply iii). Finally, since M is represented by M^o, M^1, M^2 and M^c, proposition (16.5) entails that M possesses a quadratic variation which fulfills iv). This completes the proof. □

Theorem (16.1) allows to extend the results of proposition 2.

Theorem 1: Let $M\in\mathcal{M}^2$. Then M, M^o, M^1, M^2, M^c possess quadratic variations in L^1. Moreover

i) for any enumeration $(L_n^i)_{n \in \mathbb{N}}$ of sets in $s(S^i)$ according to theorem (14.1), i)-iv) resp. theorem (14.2), i)-iv), $T(L_n^i)M$ possesses a quadratic variation in L^1 for $n \in \mathbb{N}$ and

$$[M^i] = \sum_{n \in \mathbb{N}} [T(L_n^i)M] \quad , \ i=0,1,2,$$

ii) $[M^c] = \langle M^c \rangle$ is continuous, $[M^i]$ has no 0-jumps and no $\bar{i}$-jumps, $i=1,2$,

iii) M^o, M^1, M^2, M^c have pairwise orthogonal variation,

iv) $[M] = [M^o] + [M^1] + [M^2] + [M^c]$.

Proof:

Let L be one of the sets figuring in i). For bounded martingales N, T(L)N possesses a quadratic variation which is given by one of the formulas of theorems (17.1) and (17.2). This set being dense in M^2, theorem (16.1) implies that T(L)M possesses a quadratic variation in L^1 which is given by a formula of the same kind. Also, if K is another set of i) which is disjoint to L, if it is of the same type, T(K)M and T(L)M have orthogonal variation. Now the proof of proposition 2 works with $T(L_n)$ replacing $M(L_n) - C_n$ and quadratic variations in L^1 instead of L^p, $p \geq 0$, and an appeal to theorem (19.1). □

As an immediate corollary, we obtain the following continuity theorem for quadratic variations.

Theorem 2: Let $M \in M^2$. Then:

1. if M has no 0-jumps, [M] has no 0-jumps,
2. if M has no 0-jumps and no i-jumps, [M] has no 0-jumps and no i-jumps, i=1,2,
3. if M is continuous, [M] is continuous.

Proof:

If M has no 0-jumps, $M^o = 0$. If M has no 0-jumps and no i-jumps, $M^o = M^i = 0$. If M is continuous, $M = M^c$. Therefore, the assertions follow from theorem 1. □

Remarks: Quadratic i-variations play only a peripheric role here. Let us, however, remark at this place, that our theory also yields regularity results for the quadratic i-variations of continuous square integrable martingales M. To begin with, we have $[M]^i = \langle M\rangle^i$, i=1,2. Let us briefly sketch the main facts of a proof of the continuity of $[M]^1$. For an alternative proof see Nualart [36].

Note first that $\langle M\rangle^1_{(.,1)} - \langle M\rangle_{(.,1)}$ is continuous by a well known one-parameter result (see Dellacherie, Meyer [18], p. 367) and proposition 1. Moreover, $\langle M\rangle^1 - \langle M\rangle$ is a 2-martingale, as follows from theorem (11.2). A slight extension of the definition of optional projections therefore shows

$$\langle M\rangle^1 - \langle M\rangle = {}^{\gamma}2(\langle M\rangle^1_{(.,1)} - \langle M\rangle_{(.,1)}),$$

a process which is $Q^{\pm +}$-continuous and has $Q^{\pm -}$-limits by the results of section 6. Hence $\langle M\rangle^1$ has at most 2-jumps, $\langle M\rangle$ itself being continuous. Now if M had better integrability properties, we could consider $(M^2-\langle M\rangle^1)(L)$ on an arbitrary set $L\in s(S^2)$ which is well integrable. This process is a continuous 1-martingale which, due to the continuity of M, equals $-\langle M\rangle^1(L)$ and is therefore of bounded variation. But this can only be the case if $\langle M\rangle^1(L) = 0$ (see Dellacherie [16], p. 111). Hence $\langle M\rangle^1 = [M]^1$ is continuous. If M is only square integrable, the arguments just given need to be refined a bit.

In our final theorem, we can show that quadratic variation exists even for L $\log^+$L-bounded martingales. But continuity properties are not a by-product of the proof, this time.

Theorem 3: Let M be a regular L $\log^+$L- bounded martingale. Then M possesses a quadratic variation in L^0.

Proof:

The martingales

$$M^n = E(-n \vee (M_{(1,1)} \wedge n) | \mathcal{F}.) \quad , \; n \in \mathbb{N},$$

converge to M in L $\log^+$L. They are bounded and therefore possess quadratic variations in L^0. Now we can apply the second part of theorem (16.1) to conclude. □

21. Inequalities for the quadratic variation; a counter-example

Since the beginning of multi-parameter martingale theory it has been known that L $\log^+$L is the natural boundary for the existence of regular versions. On the one hand, theorem (9.1) shows that any L $\log^+$L-bounded martingale possesses a regular version. On the other hand, Cairoli and Walsh [12] were able to exhibit an L^1-bounded martingale which is highly irregular. The main aim of this section is to show that L $\log^+$L is also the natural boundary for the existence of quadratic variations. Theorem (20.3) shows that L $\log^+$L-bounded martingales possess quadratic variations in L^0. We will modify the above mentioned example to construct an L^1-bounded martingale which has no quadratic variation in L^0. To do this, we need a martingale inequality for concave functions which compares [M] and $\sup_{t \in \Pi} |M_t|$ for martingales M and is an extension of the inequalities of Burkholder , Gundy [7].

We therefore start by investigating briefly inequalities of this kind. The convex function inequalities of theorem (8.2) extend, by simply going to the limit of the square sums along a 0-sequence of partitions of Π, to inequalities which control $[M]$ by $\sup_{t\in\Pi} |M_t|$ or even $|M_{(1,1)}|$ and vice versa.

Theorem 1: Let Φ be a moderate, Ψ a moderate and co-moderate Young function, $\Lambda(t) = t \log^+ t$, $t\geq 0$. Then there are constants $c_1, c_2, c_3 > 0$ such that for any regular $L \log^+ L$-bounded martingale M, any right-continuous martingale N on $\Omega\times[0,1]$, any $\lambda>0$

i) $c_1 \| [N]_1^{1/2} \|_\Phi \leq \| \sup_{r\in[0,1]} |N_r| \|_\Phi \leq c_2 \| [N]_1^{1/2} \|_\Phi$,

ii) $c_1 \| [M]_{(1,1)}^{1/2} \|_\Psi \leq \| \sup_{t\in\Pi} |M_t| \|_\Psi \leq c_2 \| [M]_{(1,1)}^{1/2} \|_\Psi$,

iii) $\lambda P([M]_{(1,1)}^{1/2} > \lambda) \leq c_3 \| M_{(1,1)} \|_\Lambda$.

Moreover, there is an analogue of ii) with $M_{(1,1)}$ instead of $\sup_{t\in\Pi} |M_t|$ and a corresponding analogue of i), if Φ is in addition co-moderate.

Proof:

For the one-parameter case, we refer to Dellacherie, Meyer [18], p. 304. In consequence of theorem (16.1), it is enough to consider bounded martingales. Let $(\mathcal{J}_n)_{n\in\mathbb{N}}$ be a 0-sequence of partitions of Π. We know from theorem (20.3) that

$$\sum_{J\in\mathcal{J}_n} (\Delta_J M)^2 \underset{n\to\infty}{\to} [M]_{(1,1)} \text{ in } L^0(\Omega,\mathcal{F},P),$$

at least, for any bounded martingale M. But by theorem (8.2) and the lemma of de la Vallée-Poussin, the sequence

$$(\Psi([\sum_{J\in\mathcal{J}_n} (\Delta_J M)^2]^{1/2}))_{n\in\mathbb{N}}$$

is uniformly integrable. Therefore, ii) and iii) follow from the corresponding inequalities of theorem (8.2) by going to the limit. □

Let us turn to corresponding norm inequalities for concave functions. In this case, however, our knowledge at this stage is rather poor. We are far from being able to give a "complete" set of inequalities like in the convex case. All we know at present are inequalities for the power functions $\Phi(t) = t^p$, $0<p\leq 1$. As follows from the theory of one-parameter martingales, "big jumps" of martingales are likely to destroy concave function inequalities. We will not investigate any "regularity" assumptions on the martingales or the filtration which take care of this defect (see Burkholder, Gundy [7] and Brossard [6]). Considering only continuous martingales for this reason, we will finally restrict our attention to the right half of the inequalities which seems to be the only one to be derived without too much effort. The basic idea of the following auxiliary inequality has been employed by several authors (see Brossard [6], Chevalier [14], Gundy [23]).

<u>Proposition 1</u>: For any square integrable continuous martingale M, any $\lambda > 0$

$$P(\sup_{t\in\Pi} |M_t| > \lambda) \leq 64\ P([M]^{1/2}_{(1,1)} > \lambda) + \frac{32}{\lambda^2} E([M]_{(1,1)}\ 1_{\{[M]^{1/2}_{(1,1)} \leq \lambda\}}).$$

<u>Proof</u>:

Let $\lambda>0$ and a square integrable continuous martingale M be given. We know $[M] = \langle M\rangle$ from theorem (20.1). Set

$$F = \{[M]^{1/2}_{(1,1)} \leq \lambda\}, \quad Y = {}^{\pi}1_{\overline{F}}\ , \quad X^* = \sup_{t\in\Pi} |X_t| \text{ for any process } X.$$

Obviously

(21.1) $P(M^* > \lambda) \leq P(M^* > \lambda,\ Y^* \leq 1/2) + P(Y^* > 1/2)$.

Let us first estimate the second term on the right hand side of (21.1). Theorem (10.5) gives

(21.2) $P(Y^* > 1/2) \leq 4\ \| Y^*\|_2^2 \leq 64\ P(\overline{F}) = 64\ P([M]^{1/2}_{(1,1)} > \lambda)$.

To turn to the first term, note that propositions (9.3) and (9.4) imply

$$1_{\{Y^* \leq 1/2\}}\ M = 1_{\{Y^* \leq 1/2\}}(1_{\{Y \leq 1/2\}}\cdot M.)\ .$$

Therefore

$$(21.3)\quad P(M^* > \lambda,\ Y^* \leq 1/2) = P((1_{\{Y \leq 1/2\}} \cdot M.)^* > \lambda,\ Y^* \leq 1/2)$$

$$\leq P((1_{\{Y \leq 1/2\}} \cdot M.)^* > \lambda) \leq \frac{1}{\lambda^2} \| (1_{\{Y \leq 1/2\}} \cdot M.)^* \|_2^2$$

$$\leq \frac{16}{\lambda^2} \int_{\Omega \times \mathbb{I}} 1_{\{Y \leq 1/2\}} dm_{<M>} \qquad \text{(proposition (9.3))}$$

$$\leq \frac{32}{\lambda^2} \int_{\Omega \times \mathbb{I}} (1-Y)\ dm_{<M>}$$

$$= \frac{32}{\lambda^2} E(1_F <M>_{(1,1)}) \qquad \text{(theorem (10.2))}$$

$$= \frac{32}{\lambda^2} E([M]_{(1,1)}\ 1_{\{[M]_{(1,1)}^{1/2} \leq \lambda\}}).$$

Substituting (21.2) and (21.3) in (21.1) gives the desired inequality. □

Proposition 1 readily yields an inequality for the domination of $\sup_{t \in \mathbb{I}} |M_t|$ by $[M]_{(1,1)}^{1/2}$.

Theorem 2: Let M be a square integrable continuous martingale, $0<p\leq 1$. Then

$$\| \sup_{t \in \mathbb{I}} |M_t| \|_p \leq (64 + \frac{32\,p}{2-p}) \| [M]_{(1,1)}^{1/2} \|_p .$$

Proof:

Multiply each side of the inequality of proposition 1 by $p\,\lambda^{p-1}$ and integrate with respect to λ over $\mathbb{R}_+$. □

Remarks: 1. The only proof known to us for the reverse inequality of theorem 2 is considerably more involved and requires some more information on two-parameter stochastic integration. It uses an Itô-type formula which is out of our scope here and works under the seemingly unnatural hypothesis (see section 19) that any square integrable continuous martingale can be approximated by a sequence of bounded continuous martingales. It should also work if this sequence consists of only p-integrable martingales for all p, i.e. in general, as follows from the results of section 19. See Chevalier [14].

2. Unlike in the theory of one-parameter processes, there are two more martingale functions whose norms can be compared to the

norms of $\sup_{t \in \mathbb{I}} |M_t|$ or $[M]^{1/2}_{(1,1)}$ in the style of theorems 1 and 2 and remark 1: $\sup_{t_{\overline{i}} \in [0,1]} ([M]^i_{(1,t_{\overline{i}})})^{1/2}$, $i=1,2$. Nualart [38] has inequalities linking all four functions in the convex case, for continuous martingales. We conjecture that they generalize to regular martingales. He is also able to compare $[M]^{1/2}_{(1,1)}$ to $\sup_{t_{\overline{i}} \in [0,1]} ([M]^i_{(1,t_{\overline{i}})})^{1/2}$ for a much larger class of "tame" functions which in particular includes concave ones. To do this, he essentially uses one-parameter methods developed by Lenglart, Lépingle, Pratelli [25] which apparently do not apply to $\sup_{t \in \mathbb{I}} |M_t|$.

With the support of theorem 2 we will now construct an L^1-boun-ded martingale which possesses no quadratic variation. Its prototype is a martingale figuring in a counterexample of Cairoli, Walsh [12], pp. 137, 138, which we will simply adopt.

Example: Let $(\Omega, \mathbf{F}, P)$ be a probability space on which a two-para-meter Wiener process W lives, $(\mathbf{F}_t)_{t \in \mathbb{R}^2_+}$ its completed filtration. On this space, following Cairoli, Walsh [12], p. 137, for example, we can design two sequences $(X^n)_{n \in \mathbb{N}}$, $(Y^n)_{n \in \mathbb{N}}$ of two (spa-ce) - dimensional Brownian motions, indexed by $\mathbb{R}_+$, such that

$\{X^n, Y^n : n \in \mathbb{N}\}$ is independent.

Let D denote the unit disc in $\mathbb{R}^2$, $f: \partial D \times \partial D \to \mathbb{R}$ be an integrable function whose integral is zero, h the biharmonic function on $D \times D$ whose boundary values are given by f. For $n \in \mathbb{N}$ let

$$\sigma_n = \inf \{r \in \mathbb{R}_+ : |X^n_r| = 1\} \wedge 1, \quad \tau_n = \inf \{r \in \mathbb{R}_+ : |Y^n_r| = 1\} \wedge 1.$$

Then

$$M^n_t = E(h(X^n_{\sigma_n}, Y^n_{\tau_n}) | \mathbf{F}_t), \quad t \in \mathbb{I}, \quad n \in \mathbb{N},$$

defines a sequence of independent L^1-bounded martingales. Now f can be chosen such that

$$(21.4)\quad \limsup_{t_1\uparrow\sigma_n,t_2\uparrow\tau_n} |M^n_t| = \infty \quad \text{a.s. on } \{\sigma_n<1,\ \tau_n<1\},\ n\in\mathbb{N}.$$

following Cairoli, Walsh [12], p. 137. According to this construction, M^n has one "singular point", the time when (X^n,Y^n) arrives on the boundary of $D\times D$. If we avoid touching the boundary, regularity is gained. For $n,m\in\mathbb{N}$ let therefore

$$\sigma_{m,n} = \inf\{r\in\mathbb{R}_+ : |X^n_r|=1-\tfrac{1}{m}\}\wedge 1,\quad \tau_{m,n} = \inf\{r\in\mathbb{R}_+ : |Y^n_r|=1-\tfrac{1}{m}\}\wedge 1,$$

and correspondingly

$$M^{m,n}_t = E(h(X^n_{\sigma_{m,n}},Y^n_{\tau_{m,n}})|\mathcal{F}_t)\ ,\ t\in\Pi\ .$$

Since h is biharmonic on $D\times D$, $M^{m,n}$ is a bounded continuous martingale for $n,m\in\mathbb{N}$. Moreover, (21.4) and theorem 2 give

$$(21.5)\quad \lim_{m\to\infty} E([M^{m,n}]^p_{(1,1)}) = \infty \quad \text{for } n\in\mathbb{N},\ 0<p<\infty.$$

Now the sequence $([M^{m,n}]_{(1,1)})_{n\in\mathbb{N}}$ is i.i.d. for any $m\in\mathbb{N}$. Therefore, also the sequence $(\xi^n)_{n\in\mathbb{N}}$ is i.i.d., if

$$\xi^n = \sup_{m\in\mathbb{N}} [M^{m,n}]_{(1,1)}.$$

By (21.5) and monotone convergence, ξ^n is not p-integrable for any $n\in\mathbb{N}$, $0<p<\infty$. In particular,

$$\sum_{n\in\mathbb{N}} P(\xi^n \geq n^4) = \infty.$$

Choose a sequence $(m_n)_{n\in\mathbb{N}}$ such that still

$$(21.6)\quad \sum_{n\in\mathbb{N}} P([M^{m_n,n}]_{(1,1)} \geq n^4) = \infty.$$

This is guaranteed by monotone convergence. Set

$$M = \sum_{n\in\mathbb{N}} \frac{1}{n^2} M^{m_n,n}.$$

Obviously, M is an L^1-bounded martingale. We propose to prove that M possesses no quadratic variation in L^0. For this purpose, let $(\Pi_n)_{n\in\mathbb{N}}$ be an arbitrary 0-sequence of partitions of Π. Note first that in consequence of independence, the sequence $(M^{m_n,n})_{n\in\mathbb{N}}$ has pairwise orthogonal variation. Assume that

$$[M]_{(1,1)} = \lim_{n\to\infty} \sum_{J\in\Pi_n} (\Delta_J M)^2 \quad \text{exists in } L^0(\Omega,\mathcal{F},P).$$

Then for any $k \in \mathbb{N}$ we have

$$[M]_{(1,1)} = \lim_{n\to\infty} \sum_{J\in \mathbb{I}_n} (\Delta_J(\sum_{l>k} \frac{1}{l^2} M^{m_l,l}))^2 + \sum_{l\le k} \frac{1}{l^4} [M^{m_l,l}]_{(1,1)}$$

$$\ge \sum_{l\le k} \frac{1}{l^4} [M^{m_l,l}]_{(1,1)} .$$

Consequently

$$(21.7) \quad [M]_{(1,1)} \ge \sum_{l\in\mathbb{N}} \frac{1}{l^4} [M^{m_l,l}]_{(1,1)}.$$

By independence again, the lemma of Borel-Cantelli and (21.6), (21.7) imply

$$[M]_{(1,1)} = \infty \quad \text{a.s.}$$

Therefore, M has no quadratic variation in L^0.

We remark that, what has been shown for $t=(1,1)$, could have been done for any $t \in \mathbb{I} \setminus \partial \mathbb{R}_+^2$ with the same arguments.

Remark: The example also shows that the weak Burkholder inequality of the theory of one-parameter martingales which states that there is a constant c such that for any martingale M, indexed by $[0,1]$, any $\lambda>0$,

$$\lambda P([M]_1^{1/2} > \lambda) \le c \; \| M_1 \|_1$$

(see Burkholder [9]) does not hold for two-parameter martingales. Instead, we have the $L \log^+ L$-inequality of theorem 1.

References

[1] Bakry, D. (1979). Sur la régularité des trajectoires des martingales à deux indices.
Z. Wahrscheinlichkeitstheorie verw. Gebiete 50, 149-157

[2] Bakry, D. (1981). Théorèmes de section et de projection pour les processus à deux indices.
Z. Wahrscheinlichkeitstheorie verw. Gebiete 55, 55-71

[3] Bakry, D. (1981). Limites quadrantales des martingales.
in: Processus aléatoires à deux indices,
Lecture Notes in Math. 863, 40-49. Springer, Berlin

[4] Bakry, D. (1980). Une remarque sur les semimartingales à deux indices.
Séminaire de Probabilités XV, Lecture Notes in Math. 850, 671-672. Springer, Berlin

[5] Brennan, M. D. (1979). Planar Semi-Martingales.
J. Multiv. Anal. 9, 131-151

[6] Brossard, J. (1981). Régularité des martingales à deux indices et inégalités de normes.
in: Processus aléatoires à deux indices,
Lecture Notes in Math. 863, 91-121. Springer, Berlin

[7] Burkholder, D. L. , Gundy, R. F. (1970). Extrapolation and Interpolation of Quasi-Linear Operators on Martingales.
Acta Math. 124, 249-304

[8] Burkholder, D. L., Davis, B. J., Gundy, R. F. (1972). Integral Inequalities for Convex Functions of Operators on Martingales.
Proc. Sixth Berkeley Symp. Math. Statist. Prob. 2, 223-240

[9] Burkholder, D. L. (1973). Distribution Function Inequalities for Martingales. Ann. Probability 1, 19-42

[10] Cairoli, R. (1970). Une inégalité pour martingales à indices multiples et ses applications.
Séminaire de Probabilités IV, Lecture Notes in Math. 124, 1-27. Springer, Berlin

[11] Cairoli, R. (1971). Décomposition de processus à indices doubles.
Séminaire de Probabilités V, Lecture Notes in Math. 191, 37-57. Springer, Berlin

[12] Cairoli, R., Walsh, J. B. (1975). Stochastic Integrals in the Plane.
Acta Math. 134, 111-183

[13] Cairoli, R., Walsh, J. B. (1978). Régions d'arrêt, localisations et prolongements de martingales.
Z. Wahrscheinlichkeitstheorie verw. Gebiete 44, 279-306

[14] Chevalier, L. (1982). Martingales continues à deux paramètres.
Bull. Sc. Math.(2) 106, 19-62

[15] Clarkson, J. A., Adams, C. R. (1933). On Definitions of Bounded Variation for Functions of two Variables.
Trans. Amer. Math. Soc. 35, 824-854

[16] Dellacherie, C. (1972). Capacités et processus stochastiques.
Ergebnisse der Mathematik 67. Springer, Berlin

[17] Dellacherie, C., Meyer, P. A. (1975). Probabilités et potentiel, Ch. I-IV. Hermann, Paris

[18] Dellacherie, c., Meyer, P. A. (1980). Probabilités et potentiel, Ch. V-VIII. Hermann, Paris

[19] Doléans, C., Meyer, P. A. (1979). Un petit théorème de projection pour processus à deux indices.
Séminaire de Probabilités XIII, Lecture Notes in Math. 721, 204-215. Springer, Berlin

[20] Dozzi, M. (1981). On the Decomposition and Integration of Two-Parameter Stochastic Processes.
in: Processus aléatoires à deux indices,
Lecture Notes in Math. 863, 162-171. Springer, Berlin

[21] Föllmer, H. (1984). Von der Brownschen Bewegung zum Brownschen Blatt: einige neuere Richtungen in der Theorie der stochastischen Prozesse.
Perspectives in Math., 159-190. Birkhäuser, Basel- Boston

[22] Frangos, N., Imkeller, P. (1987). Quadratic Variation for a Class of $L\log^{+}L$-bounded Two-Parameter Martingales.
Ann. Probability 15, 1097-1111

[23] Gundy, R. F. (1980). Inégalités pour martingales à un et deux indices: l'espace H^{p}.
in: Ecole d'été de Probabilités de Saint-Flour VIII-1978,
Lecture Notes in Math. 774, 251-334. Springer, Berlin

[24] Ikeda, N., Watanabe, S. (1981). Stochastic Differential Equations and Diffusion Processes. North Holland/Kodansha, Amsterdam

[25] Lenglart, E., Lépingle, D., Pratelli, M. (1979). Présentation unifiée de certaines inégalités de la théorie des martingales.
Séminaire de Probabilités XIV, Lecture Notes in Math. 784, 26-48. Springer, Berlin

[26] Mazziotto,G., Szpirglas, J. (1981). Sur les discontinuités des processus cad-lag à deux indices.
in: Processus aléatoires à deux indices,
Lecture Notes in Math. 863, 84-90. Springer, Berlin

[27] Mazziotto, G., Merzbach, E., Szpirglas, J. (1981). Discontinuités d'un processus croissant et martingales à variation intégrable.

in: Processus alèatoires à deux indices,
Lecture Notes in Math. 863, 59-83. Springer, Berlin

[28] Merzbach, E. (1979). Processus stochastiques à indices partiellement ordonnés.
Rapport interne 55, Ecole Polytechnique, Palaiseau

[29] Merzbach, E., Zakai, M. (1980). Predictable and Dual Predictable Projections of Two-Parameter Stochastic Processes.
Z. Wahrscheinlichkeitstheorie verw. Gebiete 53, 263-269

[30] Métivier, M., Pellaumail, J. (1980). Stochastic Integration.
J. Wiley, New York

[31] Métivier, M. (1982). Semimartingales. A Course on Stochastic Processes. W. de Gruyter, Berlin, New York

[32] Meyer, P. A. (1978). Une remarque sur le calcul stochastique dépendant d'un paramètre.
Séminaire de Probabilités XIII, Lecture Notes in Math. 721, 199-203. Springer, Berlin

[33] Meyer, P. A. (1981). Théorie élémentaire des processus à deux indices.
in: Processus aléatoires à deux indices,
Lecture Notes in Math. 863, 1-39. Springer, Berlin

[34] Millet, A., Sucheston, L. (1981). On Regularity of Multiparameter Amarts and Martingales.
Z. Wahrscheinlichkeitstheorie verw. Gebiete 56, 21-45

[35] Neveu, J. (1972). Martingales à temps discret.
Masson et Cie, Paris

[36] Nualart, D. (1984). On the Quadratic Variation of Two-Parameter Continuous Martingales.
Ann. Probability 12, 445-457

[37] Nualart, D. (1984). Une formule d'Itô pour les martingales continues à deux indices et quelques applications.

Introduction

There are many fields of stochastics where multi-parameter processes can be encountered. For example, to register the positions of the spins of a ferromagnetic substance at a fixed instant of time, one has to attach an appropriate state space to every point of a three-dimensional lattice. Mathematically this leads to a family of random variables indexed by a subset of $\mathbb{R}^3$, a special case of a so-called stochastic field. Correspondingly, formalizing "multivariate observations" may lead to a stochastic process indexed by a set which, according to its order properties, can be interpreted as a multi-time. The infinite dimensional Ornstein - Uhlenbeck process which appears in a variant of Malliavin's variational calculus, may be considered as a stochastic process with a multi-time or a kind of mixed space-time parameter set (see Ikeda, Watanabe [24]). A close relative of it is the "Wiener sheet" which is undoubtedly the most frequently studied among all multi-parameter processes with a continuous parameter set (see Föllmer [21]). Walsh [43] encounters the Wiener sheet in the study of mathematical models which may arise in neurophysiology or also in problems related to heat conduction or electrical cables. We finish our selection with a more recent example. The investigation of the "Poisson chaos" seems to produce a new kind of infinite dimensional Ornstein - Uhlenbeck process which, considered as a stochastic process indexed by a two-dimensional continuous variable, behaves like a Poisson process in one direction and like a Brownian motion or a more general Gaussian process in the other direction (see Ruiz de Chavez [39], Surgailis [41], [42]).

Index of definitions

Index of special symbols